TECH TRENDS IN PRACTICE

TECH TRENDS IN PRACTICE

THE 25 TECHNOLOGIES THAT ARE DRIVING THE 4TH INDUSTRIAL REVOLUTION

BERNARD MARR

WILEY

A catalogue record for this book is available from the Library of Congress.

A catalogue record for this book is available from the British Library.

ISBN 978-1-119-64619-8 (hardback) ISBN 978-1-119-64621-1 (ePDF)
ISBN 978-1-119-64620-4 (epub)

Cover design: Wiley
Cover image: © Jakarin2521/Getty Images

Set in 11/14pt MinionPro by Aptara, New Delhi, India
Printed in Great Britain by TJ International Ltd, Padstow, Cornwall, UK

10 9 8 7 6 5 4 3 2 1

To my wife Claire, and my children Sophia, James, and Oliver;
and everyone who will use these amazing technologies
to make the world a better place.

CONTENTS

CONTENTS

INTRODUCTION

We have never lived in a time of faster and more transformative technological innovation. Incredible technologies like artificial intelligence, blockchains, smart robots, self-driving cars, 3D printing, and advanced genomics, together with the other tech trends covered in this book, have ushered in a new industrial revolution. Similarly to how steam, electricity, and computers have respectively been the driving forces of the first three industrial revolutions, this fourth industrial revolution is driven by the 25 technologies featured in this book. And as with the previous industrial revolutions, this fourth industrial revolution will change businesses, reshape business models, and transform entire industries. These technologies will change how we run our businesses, what jobs we will do, and many other aspects of how we function as a society.

For most leaders it can be very challenging to keep up with the speed at which many of these new technologies are emerging. As a futurist and strategic advisor to many of the most innovative companies and governments in the world, it is my job to help leadership teams understand and prepare for the impact of these technologies. With this book, I want to provide an easy-to-understand, state-of-the-art overview of the key technologies underpinning this fourth industrial revolution and outline how they are practically used by businesses today, as well as provide some tips on how to best prepare yourself and your organization for the transformation they bring.

I have chosen these 25 technology trends because I believe that they are the key ones every business leader needs to be aware of today. There are some technologies in this book that are more foundational – like big data, 5G, and artificial intelligence – and then there are others that overlap with or use technologies like big data, 5G, and artificial intelligence – like self-driving cars, chatbots, or computer vision. My aim is to discuss the key technologies and applications that are having the biggest impact on businesses today and the medium-term future.

Before you dive into the various future tech trends, I just want to say that the fourth industrial revolution offers us huge opportunities to make our world a better place and use these technologies to address some of the world's biggest challenges – from climate change, to inequality, and from hunger to healthcare. We shouldn't waste them.

As with any new technologies, there is also huge scope to exploit them for evil and we have to put in place safeguards to ensure that doesn't happen. What is sure is that all these technologies will change businesses, reshape business models, and transform entire industries.

With many of the technologies featured in this book the rate of innovation and development is simply mind-boggling. Every week there are new breakthroughs and new applications even I didn't think possible just a few years ago. It is my job to keep a close eye on all this and I share my insights in my *Forbes* articles, YouTube videos, and across my social media channels. I would like to invite you to connect with me on LinkedIn, YouTube, Instagram, Twitter, and Facebook. I also have a weekly newsletter in which I share all the latest developments. If you would like to keep up to date then you can sign up to the newsletter on my website www.bernardmarr.com, where you can also find many more articles, videos, and reports on future tech trends.

TREND 1
ARTIFICIAL INTELLIGENCE AND MACHINE LEARNING

The One-Sentence Definition

Artificial intelligence (AI) and machine learning refers to the ability of machines to learn and act intelligently – meaning they can make decisions, carry out tasks, and even predict future outcomes based on what they learn from data.

What Is Artificial Intelligence and Machine Learning?

Speaking in 2016, Stephen Hawking said, "Success in creating AI would be the biggest event in human history." Now, it's no secret that technology trends often create a lot of hype. But in the case of AI, the hype is warranted. Like Hawking, I believe AI will transform our world and how we live in it.

AI and machine learning already plays a bigger role in everyday life than you might imagine. Alexa, Siri, Amazon's product recommendations, Netflix's and Spotify's personalized recommendations, every Google search you make, security checks for fraudulent credit card purchases, dating apps, fitness trackers … all are driven by AI.

AI and machine learning is the foundation on which many other technology trends in this book are built. For instance, without AI, we wouldn't have achieved the amazing advances in the Internet of Things (IoT, Trend 2), virtual reality (Trend 8), chatbots (Trend 11), facial recognition (Trend 12), robotics and automation (Trend 13), or self-driving cars (Trend 14) – to name just a few.

But what exactly is AI and machine learning, and how does it work? In very simple terms, AI involves applying an algorithm (a rule or calculation) to data in order to solve problems, identify patterns, decide what to do next, and maybe even predict future outcomes. Crucial to this process is an ability to learn from data and get better at interpreting data over time. And this is where the machine learning part comes in. Machine learning is a subdiscipline of AI, and it involves creating machines that can learn. ("Machines," by the way, may include computers, smart phones, software, industrial equipment, robots, vehicles, etc.)

The human brain learns from data, not a preprogrammed set of rules. We humans are continually interpreting and learning from the world around us. We generally get better at this process over time, learning from our successes and failures. And we make decisions or take action based on what we've learned. AI – or, more specifically, machine learning – replicates this process, but in machines. So, rather than just giving a machine a set of rules to follow, machines can now "learn" from data. Deep learning is another AI-related term that you might have heard. If machine learning is a subset of AI, deep learning is a subset of machine learning – it's essentially a more cutting-edge form of machine learning, involving more complex layers of data processing. (For the purposes of this chapter, both machine learning and deep learning will be wrapped up in the umbrella term AI.)

Like humans, the more data a machine has to learn from, the smarter it becomes. That explains why AI has made such dramatic advances in the last few years – advances that we might not have thought possible 10 or even five years ago. Modern AI needs data to function. And we're now creating more data than ever before (see Big Data, Trend 4). This continual expansion in data, along with advances in computing power, is fueling a rapid acceleration of AI capabilities.

AI isn't just infiltrating our everyday lives; it's going to transform our industries and businesses. According to one survey, 73% of senior executives see AI, machine learning, and automation as important areas to maintain or increase investment in.[1] (Governments, too, are prioritizing AI investment. In 2019, the White House launched a National AI Initiative directing government agencies to commit to advancing AI.[2])

As well as transforming entire businesses and industries, AI is also going to transform many human jobs. IBM predicts that more than 120 million workers globally will need to be retrained in the next three years due to AI.[3] AI-enabled automation (see Trend 22) will have a particularly significant impact and may lead to the displacement of many jobs. But rather than subscribe to a vision of a dystopian future where all human jobs are given over to robots, I believe AI will make our working lives better. Yes, jobs will be impacted by automation and AI, and many current human jobs will no longer exist in 10 or 20 years' time. But AI will enhance the work of humans, and new jobs will arise to replace displaced jobs. (Just think how computing and the internet led to the demise of some jobs but created many more new roles.) What's more, as machines become more intelligent and capable of carrying out more human tasks, I believe that our uniquely human capabilities – things like creativity, empathy, and critical thinking – will become all the more precious and valuable in the workplaces of the future.

How Is Artificial Intelligence and Machine Learning Used in Practice?

AI gives machines the ability to carry out a wide range of human-like processes, such as seeing (think facial recognition), writing (think chatbots), and speaking (think Alexa). And as machines' ability to act intelligently gets better and better, AI will infiltrate even more of our lives.

Because AI underpins so many other technology trends, throughout this book you'll find lots of specific examples of how AI is used across different businesses and industries. Here, I want to briefly whet your appetite and set out just a few of the amazing things AI can already do.

Thanks to AI, Machines Can Beat Humans at Games

Machines battling man is the theme of many a sci-fi movie. In real life, AI research and development has seen intelligent machines beat their human opponents in some significant (but thankfully less harmful) ways.

- In 1997, IBM's **Deep Blue** chess-playing machine beat world champion Garry Kasparov.[4] Many hailed this as the start of machine intelligence catching up to human intelligence, but the reality is perhaps a little less compelling. Deep Blue used brute force computing power to consider every possible chess move, and that's how it beat Kasparov. (Discover how machines learned to get a lot more creative at game-playing in Chapter 17.)

- In 2011, **IBM's Watson** AI system beat two human contestants at the game show *Jeopardy!*.[5] And not just any contestants – two of the most successful contestants the show had ever seen, who had won $5 million between them.

- In 2018, **DeepMind's AlphaStar** AI beat one of the world's most successful professional players of the real-time strategy game StarCraft II, in an impressive 5–0 victory.[6]

- In 2019, Microsoft revealed its **Suphx** AI can now beat top players at the complex Chinese tile game Mahjong. After 5,000 games, the AI was able to compete at 10th dan ranking (basically, super-expert level) – a feat that only 180 humans have ever been able to manage.[7]

- Also in 2019, we learned that AI can now solve a Rubik's Cube in 1.2 seconds – that's two seconds faster than the current world record holder (and about 20 years faster than the average human).[8] The **DeepCubeA** system was created by researchers at the University of California, Irvine.

AI Is Driving Advances in Healthcare

In August 2019 the UK government announced it would dedicate £250 million to funding AI in the National Health Service.[9] Here are just a few examples of how AI is beginning to transform healthcare.

- A study published in *The Lancet* in 2019 found that AI is as good as human experts when it comes to **diagnosing disease** from medical images.[10] Deep learning is showing enormous promise in diagnosing a range of diseases, including cancer and eye conditions.

- **MIT** researchers have developed an AI model that can predict the development of breast cancer up to five years in advance.[11] Crucially, the system works as well for black and white patients, whereas similar projects in the past have often been based overwhelmingly on white patients.

- **Infervision's** image recognition technology uses AI to look for signs of lung cancer in patient scans. The technology is already in use in healthcare settings across China.[12]

Books, Music, and Food: How AI Is Transforming Some of Our Favorite Pastimes

Content platforms like Netflix and Spotify are built on AI – they use AI to understand what viewers most want to watch or listen to, make personalized recommendations, and (in Netflix's case) create new content based on what it knows users enjoy. Here are a few other examples of AI infiltrating our hobbies and downtime.

- Chinese search engine **Sogou** has confirmed it's creating an AI that can read novels aloud, simulating the voice of their authors[13] (in a similar way to how deepfakes can create realistic audio and video content of people in the public eye). This could revolutionize audiobooks, particularly for self-published authors who perhaps don't have the means to create their own audiobooks.

- **Sony** has created an AI that can produce drumbeats for songs. Called AI DrumNet, the system was trained using hundreds of songs, and can now produce its own basic drumbeats to match other instruments on a track.[14]

- **MIT** researchers have taught AI how to reverse engineer pizza. After looking at a picture of a pizza, the AI can identify its toppings, and then tell you how to make it.[15] Why do this, you might be wondering? In theory, this technology could be used to analyze any photo of food and produce a suitable recipe. So if you want to recreate an amazing restaurant meal at home, in a few years' time there might be an app for that!

The Future of AI?

In 2019, Microsoft announced it was plowing $1 billion into AI research lab **OpenAI** – which was founded by, among others, Elon Musk.[16] What's behind such a big investment? OpenAI is dedicated to creating something called artificial general intelligence (AGI), widely considered to be the "holy grail" of AI.

While AI can do some incredible things when it comes to "general intelligence," AI lags way behind the human brain. In other words, AI is great at learning to do specific things, but AI systems can't just apply that knowledge to other tasks in the way humans can. This is the goal of AGI – to create an AI system that's as generally intelligent and flexible as the human brain. It's not been done yet – in fact, we don't know if AGI is even possible – but Microsoft's investment shows it's certainly a serious goal.

Key Challenges

I opened this chapter with a Stephen Hawking quote, "Success in creating AI would be the biggest event in human history." Hawking immediately followed that up with, "Unfortunately, it might also be the last, unless we learn how to avoid the risks."

AI isn't without its challenges and risks. For one thing, there are potentially huge risks for society and human life as we know it (particularly when you consider some countries are racing to develop AI-enabled autonomous weapons). But let's focus on the key challenges that everyday businesses will have to overcome if they're to deploy AI successfully.

Regulation

There will no doubt be regulatory hurdles to negotiate as regulators begin (quite rightly and belatedly) to take a greater interest in the application of AI. Until now, some of the early adopters of AI have played a bit fast and loose with the technology (Facebook, for example, is facing legal action over its use of facial recognition technology for auto-tagging photos, without gaining user consent).[17] That sort of behavior can't continue, and business leaders will have to take an ethical, responsible approach to AI.

Privacy Concerns

Part of using AI ethically means making sure you respect individuals' privacy, gain consent to use their data for AI applications, and make it clear how you are using their data. Again, this is where some big players have fallen short in the past. Amazon, for example, faced consumer outrage over the news that contractors were listening to people's Alexa requests. Individuals could not be identified by their audio, and Amazon stressed the practice was necessary to help develop Alexa's capabilities, but the fact remains that most users had no idea that anyone would ever hear their private audio. Amazon has since introduced a "no human review" option to its Alexa settings, which allows users to opt out of their audio being manually reviewed.[18]

Lack of Explainability

Remember I said that AI can now solve a Rubik's Cube in just 1.2 seconds? Interestingly, the researchers who built the puzzle-solving AI can't quite tell how the system did it. This is known as the "blackbox problem" – which means, to put it bluntly, we can't always tell how very complex AI systems arrive at their decisions.

This raises some serious questions around accountability and trust. For example, if a doctor alters a patient's treatment plan based on an AI prediction – when he or she has no idea how the system arrived at that prediction – then who is responsible if the AI turns out to be wrong? What's more, under GDPR (the General Data Protection Regulation legislation brought in by the European Union), individuals have the right to obtain an explanation of how automated systems make decisions that affect them.[19] But, with many AIs, we simply *can't* explain how the system makes decisions.

New approaches and tools are currently being developed that help to better understand how AIs make decisions but many of these are still in their infancy.

Data Issues

Put simply, AI is only as good as the data it's trained with. If that data is biased or unreliable, then the results will be biased or unreliable.

For example, facial recognition technology was found to be generally better at identifying white males than women and people of color, because a leading data set used to train facial recognition systems was estimated to be more than 75% male and 80% white – something that programmers were able to correct by adding a more diverse range of faces to the training dataset.[20] This means companies will need to ensure their data is as unbiased, inclusive, and representative as possible if they're to get the best results from AI.

The AI Skills Gap

Finally, one area in which many companies will struggle is finding the right AI talent. There's a shortage of people who can develop these complex AI systems – and what talent there is tends to be scooped up by the Googles and IBMs of this world. AI-as-a-service (AIaaS) could be part of the solution. AIaaS offerings from companies like IBM and Amazon allow companies to make use of AI tools, without having to invest in expensive infrastructure or new hires, which makes AI much more accessible to businesses of all shapes and sizes.

How to Prepare for This Trend

AI is going to revolutionize almost every facet of modern life, including business. Therefore, despite the challenges involved, businesses cannot afford to overlook the potential of AI. So how might you use AI in your business? Broadly speaking, companies are using AI to improve their business in three ways:

- Developing smarter products (see Trends 2 and 3 for great examples of this).

- Delivering smarter services (check out Trends 18 and 23 as examples of AI-driven services).

- Making business process more intelligent (Trends 12, 13, and 17 for just a few examples of AI-enhanced business processes).

Every business should consider whether they can use AI to improve their business in one or, ideally, all of these ways. But you'll need a robust AI strategy in order to get the most out of AI – and a good AI strategy should always be linked to your overarching business strategy. To put it another way, you need to look at what the business is trying to achieve and then see how AI can help you deliver those strategic goals.

Notes

1. 7 Indicators Of The State-Of-Artificial Intelligence (AI), March 2019, *Forbes*: www.forbes.com/sites/gilpress/2019/04/03/7-indicators-of-the-state-of-artificial-intelligence-ai-march-2019/#5d371cbb435a
2. White House Unveils a National Artificial Intelligence Initiative: www.nextgov.com/emerging-tech/2019/02/white-house-unveils-national-artificial-intelligence-initiative/154795/
3. More Robots Mean 120 Million Workers Will Need to be Retrained, *Bloomberg*: www.bloomberg.com/news/articles/2019-09-06/robots-displacing-jobs-means-120-million-workers-need-retraining
4. How Did A Computer Beat A Chess Grandmaster?: www.sciencefriday.com/articles/how-did-ibms-deep-blue-beat-a-chess-grandmaster/
5. Watson and the Jeopardy! Challenge: www.youtube.com/watch?v=P18EdAKuC1U
6. AlphaStar: Mastering the Real-Time Strategy Game StarCraft: https://deepmind.com/blog/article/alphastar-mastering-real-time-strategy-game-starcraft-ii
7. After 5,000 games, Microsoft's Suphx AI can defeat top Mahjong: https://venturebeat.com/2019/08/30/after-5000-games-microsofts-suphx-ai-can-defeat-top-mahjong-players/
8. AI learns to solve a Rubik's Cube in 1.2 seconds: www.engadget.com/2019/07/17/ai-rubiks-cube-machine-learning-neural-network/
9. Boris Johnson pledges £250m for NHS artificial intelligence, *The Guardian*: www.theguardian.com/society/2019/aug/08/boris-johnson-pledges-250m-for-nhs-artificial-intelligence
10. A comparison of deep learning performance against health-care professionals in detecting diseases from medical imaging, *The Lancet*: www.thelancet.com/journals/landig/article/PIIS2589-7500(19)30123-2/fulltext

11. MIT AI tool can predict breast cancer up to 5 years early, works equally well for white and black patients: https://techcrunch.com/2019/06/26/mit-ai-tool-can-predict-breast-cancer-up-to-5-years-early-works-equally-well-for-white-and-black-patients/
12. Infervision: Using AI and Deep Learning to Diagnose Cancer: www.bernardmarr.com/default.asp?contentID=1269
13. The Search Engine AI That Reads Your Books: www.aidaily.co.uk/articles/the-search-engine-ai-that-reads-your-books
14. Sony's new AI drummer could write beats for your band: https://futurism.com/the-byte/sony-ai-drummer-write-beats-your-band
15. MIT's new AI can look at a pizza, and tell you how to make it: https://futurism.com/the-byte/mit-pizza-ai
16. Microsoft invests $1 billion in OpenAI to pursue holy grail of artificial intelligence: www.theverge.com/2019/7/22/20703578/microsoft-openai-investment-partnership-1-billion-azure-artificial-general-intelligence-agi
17. Facebook faces legal fight over facial recognition: www.bbc.com/news/technology-49291661
18. Amazon quietly adds "no human review" option to Alexa settings as voice AIs face privacy scrutiny: https://techcrunch.com/2019/08/03/amazon-quietly-adds-no-human-review-option-to-alexa-as-voice-ais-face-privacy-scrutiny/
19. The "right to an explanation" under EU data protection law, *Medium*: https://medium.com/golden-data/what-rights-related-to-automated-decision-making-do-individuals-have-under-eu-data-protection-law-76f70370fcd0
20. How Bias Distorts AI, *Forbes*: www.forbes.com/sites/tomtaulli/2019/08/04/bias-the-silent-killer-of-ai-artificial-intelligence/#260abf2e7d87

TREND 2
INTERNET OF THINGS AND THE RISE OF SMART DEVICES

The One-Sentence Definition

The Internet of Things (IoT) refers to the increasing number of everyday devices and objects that are connected to the internet and are capable of gathering and transmitting data.

What Is the Internet of Things?

The rise of smart devices has played a key role in the massive explosion of data (see Big Data, Trend 4) – and is rapidly changing our world and the way we live in it. But, in the IoT, data is created by things, not people, which has given rise to the term "machine-generated data." How exactly are machines generating data? Typically, it's when smart devices, gadgets, or machines gather information and communicate that data via the internet – an example being your fitness tracker automatically sending activity data to an app on your phone. (However, as we'll see later in this section, in the future, devices will increasingly process the data themselves, without having to transmit it for analysis.)

This is all possible because, these days, pretty much everything is getting smarter. It all started with the iPhone, and has since snowballed to include smart TVs, smart watches, and fitness trackers (see wearables,

Trend 3), smart home thermostats, smart fridges, smart industrial machinery…even smart nappies that alert you when your baby has, well, done what babies do best. A huge range of devices, machines, and equipment are now fitted with sensors and have the ability to constantly gather and transmit data. Today, even the smallest devices can effectively function as a computer. (However, it's important to note that an actual computer wouldn't count as part of the IoT, since the IoT generally refers to everyday objects that we wouldn't traditionally expect to be able to connect to the internet – like fridges and TVs.)

An IoT device could be as small as a light bulb – smaller, in some cases – or as large as a streetlamp – see intelligent spaces and smart places, Trend 5 – and may be found at home, on our city streets, in our offices, in healthcare settings, in industrial settings, and more. I delve into some of the practical applications of the IoT later in the chapter.

The ability of machines to connect to and share information with each other is a key part of the IoT. These machine-to-machine conversations mean that devices can talk to each other and potentially decide on a course of action without human intervention. For example, manufacturing equipment fitted with sensors could transmit performance data to the cloud for analysis, and based on that data, the system could automatically schedule the equipment for repair and maintenance. (The use of the IoT in industrial and manufacturing settings is often referred to as "Industry 4.0" – smart industry, in other words.)

How big is the IoT? Pretty darn big. The IoT has experienced enormous growth in recent years and the popularity of smart devices shows no sign of slowing down. IHS predicts that 75 billion devices will be connected to the internet by 2025.[1] If that number seems hard to fathom, consider this: as of January 2019, Amazon had sold more than 100 million smart devices with Alexa installed.[2] That's just Amazon Echo smart speakers and other Alexa-enabled devices! (Interestingly, many IoT devices are catching on to the power of voice interfaces like Alexa – see Trend 11.)

As well as becoming more ubiquitous, these smart devices are also becoming more powerful, which means that more of the computing can be done on them, rather than having to upload data to the cloud for analysis. This is what's known as "edge computing" (see Trend 7). With edge computing, data is processed closer to the source of the data and away from the cloud – in theory meaning that your smart fridge could process data itself. It's relatively early days for edge computing, but it's predicted to bring big benefits. If you think about it, IoT devices create masses of data – not all of it critical – which can slow down processing and decision-making (fine if you're just asking Alexa for the weather report, but definitely not fine in, say, a self-driving vehicle). With edge computing (see also Trend 7), networks are less clogged because more processing is happening closer to the data source, which means critical data can be handled much more quickly.

Edge computing is just one of the advances we can look forward to in the IoT, but it's by no means the only one. As businesses quickly cotton on to the power of the IoT, expect many more exciting IoT-related developments in the coming years.

How Is the Internet of Things Used in Practice?

The IoT is set to become even more deeply embedded in our everyday lives: at home, at work, and when we're on the move. In fact, you might be surprised how deeply entrenched it is already. Let's look at some of my favorite real-life examples of the IoT in action.

Making Our Homes and Everyday Lives Smarter Through Intelligent Consumer Goods

Unlocking your front door with a boring old key? No need, if you have a smart front door lock. Turning light switches on and off with your actual hands? What are you, a cave dweller? The idea behind many smart consumer goods is to simplify (and even automate) those mundane, everyday tasks. What's more, the best of today's smart products get to know your preferences and behavior, so that they can anticipate your needs and respond to your behavior. For example:

- Google-owned **Nest's** learning thermostat tracks how you use your home so that it can regulate your home's temperature accordingly.

- The **Orro** intelligent light switch can tell when you're in the room and switch the lights on and off without you having to do anything. It'll also adjust the lighting based on the time of day.

- The **August Smart Lock Pro** allows you to lock and unlock your home from anywhere, without a key. It automatically locks the house when you leave and unlocks it when you come home and can integrate with voice assistants like Alexa and Siri.

- **LG's smart wine fridge** can tell you what food to pair with your tipple, and recommend which wine to buy next, based on what it learns about your tastes.

- **LINKA's** smart bike lock recognizes you as you approach and automatically unlocks your bike, without a key. You can also grant remote access to your bike to family and friends.

- You can even get smart toilets these days. No really. The **Kohler Numi 2.0 Intelligent Toilet** comes with built-in Amazon Alexa – a snip at $8,000.

Wearable devices such as smart watches, fitness trackers, and even smart clothes make up a critical part of the IoT. Read more about the wearables trend in Chapter 3.

Making People Healthier with the Internet of Medical Things (IoMT)

The IoT is poised to transform the healthcare industry, giving rise to its own name: the IoMT. These IoMT devices can be used to help monitor patients, inform caregivers in the event of an emergency, and provide healthcare professionals with data that could inform diagnosis and ensure patients follow doctors' orders. For example, IoMT devices can track vitals and heart performance, monitor glucose and other body systems, and track activity and sleep levels. Think about the impact of this for a second – instead of doctors relying on what the patient tells them, the IoMT gives healthcare providers an incredible insight into what's really going on with the patient's health and lifestyle. As you can probably imagine, the IoMT is closely linked to the rise of wearable technology (see also Trend 3).

Transforming the Way We Do Business

The IoT offers huge benefits to businesses. There are some great examples of this.

- For companies that make and sell products, making those products smart can deliver unprecedented insights into how those products are used. Thanks to these insights, companies can deliver a better service and improved products. **Rolls-Royce**, for example, installs sensors in the jet engines it manufactures, so it can better understand how airlines use those engines.

- The IoT also gives businesses the chance to deliver new customer value propositions. For instance, tractor and farm equipment manufacturer **John Deere** has developed intelligent farming solutions where sensors continuously monitor soil health and other factors, and give farmers advice on what crops to plant where, and so on.

- Companies are also generating new income streams, thanks to the IoT. **Google's Nest** smart thermostats are one example of this. The thermostats collect real-time energy usage data from customers – data that's incredibly valuable to utility companies and other interested parties. In this way, data generated from IoT devices can become a key business asset, and potentially bolster the company's value.

- For many companies, the biggest IoT opportunities lie in the ability to improve and optimize operations. Data generated from smart machines – for example, manufacturing equipment – can be used to improve the way the company is run, potentially automate various processes, drive efficiencies, improve reliability, reduce costs, and so on. It's no surprise, then, that manufacturing and industrial companies have been leading adopters of IoT technology, which brings me to…

Introducing the Industrial Internet of Things (IIoT)

Companies are increasingly seeing the value in connected machinery that is capable of reporting every detail of operations – and this network of connected industrial devices is known as the IIoT. Examples include:

- Robotics and automation company **ABB** used connected IIoT sensors to monitor its robots' maintenance needs, so it can carry out repairs and maintenance before parts break.

- Automobile parts manufacturer **Hirotec** used IIoT technology to monitor the reliability and performance of machinery at one of its tool-building operations. The data was used to make machines more productive. The company is now focusing on connecting a whole production line at one of its manufacturing plants in Japan. This means the production of a complete auto component – in this case, a car door – will all happen in a smart, connected way.[3]

- The IIoT is even helping trains run on time. **Siemens AG** gathers data from sensors on trains and rail infrastructure to, among other things, carry out predictive maintenance and increase energy efficiency. As a result, the company says it can now guarantee almost 100% reliability for its customers.[4]

- Field service management provider **ServiceMax** has created an IIoT-driven platform called Connected Field Services to help companies implement predictive maintenance for their mobile and offsite equipment. Eventually, ServiceMax hopes the platform will help guarantee 100% uptime availability for mission-critical equipment.[5]

Get Ready for Smart Dust

Wireless devices the size of a grain of salt that are equipped with tiny sensors and cameras? Already a reality. Microelectromechanical systems (MEMs, or motes, as they're sometimes called) are very real and have the potential to multiply the IoT millions or billions of times over.[6] In the future, MEMs may be used in settings like agriculture, manufacturing, and security, as well as in robotics (Trend 13) and drone technology (Trend 19).

Key Challenges

Privacy is one concern with IoT devices – how much of our activity and behavior do we really want monitored, especially in our own homes? Many people appear happy to forgo their privacy in return for a smarter, more efficient home, yet advances like smart dust – where devices are so small they're hard to detect – may make this a bigger concern going forward. For this reason, companies looking to embed IoT technology in their products, workplaces, and equipment should absolutely take privacy, ethics, and transparency seriously.

Security is another major concern. For one thing, the connectivity of IoT devices has created a dangerous side effect known as botnets. The term refers to a group of internet-connected devices controlled by a central system, and is usually related to DDoS attacks – where a hacker uses a large group of devices to flood a website with fake requests, bringing the website down. A famous example of this is the 2016 DDoS attack that took a major internet provider partially offline, causing many high-profile websites like Twitter and Amazon to disappear from the internet for a while. In that attack, an estimated 100,000 unsecured IoT devices were harnessed to create the botnet.[7]

The problem is many of these devices that connect to the internet have little or no built-in security – and even when they do, users often neglect to take basic security precautions, like setting a password. This makes the problem of botnets much worse – and also opens up devices to data theft. In other words, your smart devices could potentially leak your data, and offer easy access points to anyone looking to steal it.

Particularly for organizations – but really, for anyone with an IoT device – it's vital to take the necessary steps to secure your devices and data. As well as protecting devices with passwords, this may include:

- **Routinely ensuring devices are up to date with the latest version of software.** New weaknesses are constantly being found, and patches are released regularly to fix security gaps. Hitting the "update later" option could leave your system vulnerable to attack.

- **Auditing devices on a regular basis.** Companies increasingly allow employees to connect their own devices to company networks (what's known as "bring your own device"), but this can create security headaches. Keep a record of every device that has access to your network, including company devices, and ensure all devices are up to date with the latest operating system.

- **Segmenting networks** so that different parts of the network that don't need to talk to each other are kept isolated from each other.

Similarly, if something doesn't need to be connected to your network, then don't connect it.

- **Keeping an eye out for botnets.** Analyzing your network traffic is the best way to spot this. If you notice your devices are habitually connecting or sending data to destinations you don't recognize, they may need updating or be taken offline.

Blockchain (Trend 6) may play an increasing role in IoT security. According to one report, the use of blockchain technology to secure IoT devices and data doubled during 2018.[8] The powerful encryption used to secure blockchains makes it very hard for attackers to penetrate even one part of the chain.

How to Prepare for This Trend

Despite security threats, the IoT offers incredible opportunities for businesses looking to better understand their customers, streamline operations, create new customer value propositions, and drive revenue – providing you prepare your organization accordingly. Here are a few key steps to help prepare for this trend:

- **Consider how the IoT relates to your overarching business strategy.** For the IoT to deliver real value, it must be linked to your business goals. What is your business trying to achieve – for example, understanding customers better, reducing operating costs – and how could the IoT propel your business towards those goals?

- **If you make products, consider whether you could make those products more intelligent.** In this age of smart everything (even toilets!), customers increasingly expect their everyday goods to provide more intelligent solutions.

- **Don't overlook data storage needs.** The IoT brings with it enormous amounts of data. Do you have the storage and computing power to store and make sense of all that data?

- **Think about how you'll analyze all that data.** There are lots of off-the-peg solutions designed to help you make sense of your IoT-related data.

- **Make sure your IoT data is accessible to those who need it.** Capturing all this data is great, but it must be useful. That means various people in the company need to be able to access and interpret the data, so that they can make better decisions, streamline operations, and so on.

- **Create a clear IoT security strategy** that sets out which department is responsible for minimizing the threat of attack through connected devices, who is responsible for auditing and updating machines on the network, and what to do in the event of a breach.

Notes

1. Do you know the tenets of a truly smart home? *Wired*: www.wired.com/brandlab/2018/11/know-tenets-truly-smart-home/
2. More than 100 million Alexa devices have been sold: https://techcrunch.com/2019/01/04/more-than-100-million-alexa-devices-have-been-sold/
3. Hirotec: Transforming Manufacturing With Big Data and the Industrial Internet of Things (IIoT): www.bernardmarr.com/default.asp?contentID=1267
4. Siemens AG: Using Big Data, Analytics And Sensors To Improve Train Performance: www.bernardmarr.com/default.asp?contentID=1271
5. ServiceMax: How The Internet of Things (IoT) and Predictive Maintenance Are Redefining the Field Service Industry: www.bernardmarr.com/default.asp?contentID=1268
6. Smart Dust Is Coming. Are You Ready? *Forbes*: www.forbes.com/sites/bernardmarr/2018/09/16/smart-dust-is-coming-are-you-ready/#27afb4125e41
7. Lessons from the Dyn DDoS Attack, Schneier on Security: www.schneier.com/essays/archives/2016/11/lessons_from_the_dyn.html
8. Almost half of companies still can't detect IoT device breaches, reveals Gemalto study: www.gemalto.com/press/Pages/Almost-half-of-companies-still-can-t-detect-IoT-device-breaches-reveals-Gemalto-study.aspx

TREND 3
FROM WEARABLES TO AUGMENTED HUMANS

The One-Sentence Definition

This trend harnesses artificial intelligence (AI, Trend 1), the Internet of Things (IoT, Trend 2), Big Data (Trend 4), and robotics (Trend 13) to create wearable devices and technology that help to improve the physical – and potentially mental – performance of humans, and help us lead healthier, better lives.

What Are Wearables and Augmented Humans?

Perhaps the most prevalent examples of wearables today are fitness tracker bands and smart watches – small, easy-to-wear devices that typically monitor our activity and provide insights that help us lead healthier, better, more productive lives. However, the term "wearable" doesn't necessarily mean something that you strap onto your wrist or wear elsewhere on your body; it also extends to "smart" clothing, such as running shoes that can measure your running gait and performance, advances like robotic prosthetics, and robotic wearable technology used in industrial settings.

As technology gets smaller and smarter, the sheer range of wearables is going to expand enormously – and new, smaller, smarter products will emerge to supersede the wearables we're familiar with today. For example, we already have smart glasses, but these are likely to be replaced by smart contact lenses (see practical applications later in the chapter). And after that, smart contact lenses will likely be replaced by smart eye implants.

Advances like this lead many to believe that humans and machines will eventually merge to create truly augmented humans – "transhumans" or humans 2.0 if you like, where the human body is "souped up" like a sports car to achieve enhanced physical and mental performance. This would transform the world of medicine – some believe disabilities as we know them today won't exist in the future – and, eventually, may even challenge our understanding of what it means to be human.

Sound far-fetched? Not at all when you consider that we already have advanced robotic limbs that can replace human limbs and, thanks to AI, be controlled by the wearer's thoughts (more on this coming up). And we won't just be looking at physical augmentations, either. AI for the human brain is already in development. Companies like Facebook are racing to develop brain–computer interfaces that could, in theory, allow you to type your Facebook status update using your mind instead of your fingers (telepathic typing, to use the vaguely creepy technical term).[1] Similarly, Elon Musk's Neuralink company is working on a brain–computer interface that would help people with severe brain injuries. Musk, who has spoken openly about his concerns for the human race as machines become increasingly intelligent, believes merging with machines and enhancing our human capabilities may be the best way to stop us being wiped out by our intelligent creations, or turned into their "pets."[2]

So in the future, we may find ourselves permanently attached to our smart phones, but in a more literal way – because the technology could be implanted into our bodies and capable of constantly scanning our thoughts, emotions, and biometric data to understand what we want to do next. AI chips implanted in our brains could help us make smarter, faster decisions. And physical augmentation could make us stronger, faster, and who knows what else. No longer satisfied with manipulating the world around us, it seems humans are on a quest to manipulate *themselves*.

How Are Wearables Used in Practice?

It all started with smart watches and fitness trackers. These now commonplace wearable devices are designed to help us lead healthier lives – and research suggests it actually works. One study found that participants with an Apple Watch linked to health and life insurance reward schemes increased their activity levels by a third, potentially amounting to an extra two years' life expectancy.[3] Smart watches also now have the ability to spot heart problems; the Apple Watch Series 5 has the ability to take an ECG, recording your heartbeat and rhythm in the same way as a hospital machine would, and is considered an approved medical device by the United States' Food and Drug Administration.[4]

Soon, capabilities like this will be par for the course for all smart watches, fitness trackers, and other smart devices. But there are many other exciting (and occasionally downright weird) advances to get to grips with the world of wearables, from smart clothes, to technology that physically augments the human body, to the eventual merging of the human brain with computers.

Let's take a look at each category in turn.

Smart Clothes for Smarter Lives

Clothes are becoming more intelligent, with a view to making our lives better and more convenient. These smart clothes are otherwise regular garments that have been enhanced with technology – such as sensors or high-tech circuitry – which allows them to perform functions way beyond protecting our modesty or keeping us warm and dry. Here are some of my favorite examples of smart clothes already on the market:

- Designed for athletes and serious exercise fanatics, **Under Armour's Athlete Recovery Sleepwear** is designed to improve muscle recovery and deliver a better night's sleep by absorbing the wearer's body heat and releasing infrared light.

- **Ralph Lauren's PoloTech** t-shirts are fitted with biometric sensors that monitor heart rate and other metrics and deliver workout insights to your smart phone or watch, including tailored workout advice.

- Designed for runners, **Sensoria Smart Socks** monitor pressure on your feet while running and send data to your smart phone. (Not all smart socks are for fitness enthusiasts, however. **Siren's Diabetic Sock and Foot Monitoring System** monitors the wearer's temperature to detect early signs of inflammation, which can lead to foot ulcers in diabetics.)

- **Wearable X's Nadi yoga pants** vibrate at various points (such as the knee or hip) to encourage you to move or hold positions. By syncing with an accompanying app, the pants give additional feedback on your yoga positions.

- Fashion tech startup **Supa** has a smart bra complete with heart rate sensor and AI that tracks your workouts. Naturally, it syncs to an app so you can keep track of your health data over time.

- Looking beyond workout gear, **Tommy Hilfiger** wants you to wear its casualwear so much, it's introduced a whole line of clothing that tracks how often you wear items and gives you rewards for frequent usage. The clothing line includes hoodies, jeans, and t-shirts, all with embedded chips that send info to an accompanying app.

- **Google and Levi's** have collaborated on a smart denim jacket, called Jacquard, that connects to the wearer's smart phone. With a tap or swipe of the sleeve you can control music volume on your phone, screen calls, get directions, and receive updates on your Uber ride.

- Smart socks that monitor your baby's heart rate as it sleeps? Sounds like the perfect gift for any anxious new parent (which, let's face it, is every new parent). The **Owlet Smart Sock** not only monitors heart rate and breathing interruptions, it can also identify potential health issues like sleep irregularities, heart defects, lung disorders, or pneumonia.

Wearable Technology that Physically Augments Humans

From prosthetics that help restore amputees' motor functions to industrial equipment that helps employees work smarter and safer, wearable technology goes way beyond everyday smart watches or clever yoga pants. Let's look at some of the amazing ways wearable technology is physically enhancing the human body.

Improving Human Strength and Balance

- Exoskeletons – essentially, wearable robot suits – already exist that help workers become super-strong. For example, the **Sarcos Guardian XO exoskeleton** is a full-body suit that lets workers in, say, factory and construction settings lift up to 90 kilograms without strain. Sarcos says the technology will help to increase productivity and reduce workplace injuries. In case you're wondering what a "full-body suit" looks like, picture the get-up Ripley wore for her epic alien fight in the movie *Aliens*, and you won't be far off!

- In 2018, **Ford** confirmed it was rolling out 75 EksoVest upper-body exoskeletons in a number of its auto plants around the world – at the time of writing, the largest adoption of exoskeletons to date.[5] Volkswagen is also exploring rolling out rival exoskeletons at its plants.[6]

- There are actually many different types of exoskeletons, and not all of them designed with industrial super-strength in mind. Many are designed for clinical rehabilitation purposes, for instance, by helping to provide support to the hips, legs, and lower torso. The **Rewalk Robotics Restore soft exoskeleton**, which is designed to help stroke patients walk effectively and efficiently, is a good example.

- **MIT** has developed a robot that can understand signals from your muscles and respond accordingly, to help you lift heavy objects. The mechanical system works by reading the electrical signals from your biceps – measuring your flex, in other words – to get a sense of how you're lifting. It can then work out how best to help you lift. It might not be for the squeamish, though, since it requires electrodes to be inserted into your arm![7]

- If you don't fancy electrodes in your arm, how about a robotic tail? Designers in Japan have developed a **robo-tail** that straps around the waist and helps improve people's balance.[8] It's not commercially available, but the designers predict robo-tail technology could in future help with rehabilitation or to augment balance for workers in dangerous locations, such as on a construction site.

Improving People's Sight

- **Ocumetrics** has created a Bionic Lens that claims to give wearers vision that's three times better than what we'd normally consider perfect (i.e. 20/20) vision. The Bionic Lens comes folded up, like a taco, and is implanted into the eye in a quick, painless procedure – after which it unfolds itself over the eye in a matter of seconds, immediately correcting your sight.[9] If the lens becomes widely available, pending clinical trials, it could make glasses and regular contact lenses a thing of the past.

- Elsewhere, **Samsung** has been granted a patent for smart contact lenses that are capable of taking photos and recording video. The design also includes motion sensors, which would allow wearers to control devices with eye movements.[10] If Samsung does end up making the lenses, they could become a serious challenger for smart glasses, such as Google Glass (see augmented reality, Trend 8).

Restoring Movement Through Advanced Robotic Limbs

- Prosthetics have come a long way, and the cutting edge is prosthetics that are controlled by neural activity to restore motor function to amputees. One example is an advanced mind-controlled robotic arm created by **Johns Hopkins Applied Physics Lab.**[11]

- Developed by Haptix, DEKA, and the University of Utah, the **Luke neuroprosthetic hand** (yes, it's named after Luke Skywalker) aims to restore the sense of touch to amputees by helping recipients "feel" intuitively through the prosthetic.[12] In tests, the wearer was able to pick up an egg without cracking it and hold his wife's hand – and thanks to electrodes implanted in the subject's forearm, the hand triggered touch sensations such as vibrations, pain, pressure, and tightening.

- Elsewhere, at the **National University of Singapore**, scientists have created an artificial skin that can sense better than human nerves, and could one day be used to cover prosthetic limbs.[13]

Implanting Lab-Grown Organs

- Researchers at **Massachusetts General Hospital and Harvard Medical School** have teamed up to create stem cells that can be used to form heart tissue. The tissue even beats when given an electric shock. And at the University of Glasgow and the University of the West of Scotland, scientists have used bone marrow cells to create a putty that can be used for bone grafts.[14]

- We could even 3D print organs. For example, bioprinting company **Organovo** has been able to 3D print human liver tissue patches, which have been successfully implanted into mice[15] (more on that in Trend 24).

Augmenting the Human Brain Through Mind-Reading Technology

In the future, wearable technology may not be limited to enhancing humans' physical activity, but our mental activity, too. Here are two prominent examples of the move towards merging humans with computers:

- In a study backed by **Facebook**, scientists at the **University of California San Francisco** have created a brain–computer interface that translates brain signals into dialogue – meaning technology can decode speech directly from the human brain, without us having to say or type a word.[16]

- Never one to be outdone by Mark Zuckerberg, Elon Musk's **Neuralink** company is working towards an ultimate goal of merging the human brain with AI, with human trials potentially coming this year.[17]

Key Challenges

As the examples in this chapter show, we're clearly well on our way to augmented humans. The prospect of humans merging with machines no longer seems like the imaginative plot of a sci-fi movie – but is a genuine goal for some technology companies. But with this ambitious goal comes some major challenges.

For one thing, if projects like the mind-reading technology being developed by Facebook and Neuralink do succeed, it could have huge implications for privacy. Do we really want AI to be able to decode our thoughts? And do we really want that data in the hands of for-profit companies like Facebook? I know I don't. Before technology like this becomes the norm, there would need to be some serious leaps in people's understanding of the precious data they're giving over to these companies (considering that most people today, in my experience, massively underestimate the things companies like Facebook and Google already know about them). And the companies offering this technology would need to make genuine strides in how seriously they take data privacy and ethics.

And at a societal level, we could be heading towards even greater divides between rich and poor, between the haves and the have-nots. Technology is promising to help us live longer and healthier lives – maybe even the chance to live forever – but probably only for those who can afford it. Imagine a society in which the rich are effectively superhumans that live forever, and everyone else is enormously disadvantaged. Not a happy thought, is it? (There's also a wider ethical question around whether we should *want* to live exceedingly long lives, given the huge strain that would put on our planet.)

Finally, as humans begin to merge with machines, we may ultimately need to rethink what it means to be human. Will, for example, AIs be covered by human rights legislation? And what will the term "human rights" even mean when humans have transformed themselves into something entirely new?

How to Prepare for This Trend

Finding practical takeaways and action points can be tough when you're talking about heavy stuff like the very fabric of what makes us human! But there are practical steps your organization can take to benefit from what's happening in wearable technology *right now.*

From the plethora of smart devices and smart clothing on the market, it's clear that consumers are welcoming intelligent wearables that can deliver new insights and help them live healthier, better lives.

Therefore, if your company manufactures wearable products or devices, consider whether it's possible to make those products smarter and deliver more value to your customers through intelligent insights. And from a service point of view, consider whether the wearables trend could help you offer more intelligent services. The insurance industry provides a great example of this, where customers with health or life insurance policies are rewarded for leading healthier, more active lives, by tracking their activity data through a smart watch or fitness tracker.

Notes

1. Here Are the First Hints of How Facebook Plans to Read Your Thoughts: https://gizmodo.com/here-are-the-first-hints-of-how-facebook-plans-to-read-1818624773
2. Elon Musk Isn't the Only One Trying to Computerize Your Brain. *Wired*: www.wired.com/2017/03/elon-musks-neural-lace-really-look-like/
3. Apple Watch could add two years to your life, research suggests. *The Telegraph*: www.telegraph.co.uk/news/2018/11/28/apple-watch-could-add-two-years-life-research-suggests/
4. Apple Watch 4 is Now An FDA Class 2 Medical Device. *Forbes*: www.forbes.com/sites/jeanbaptiste/2018/09/14/apple-watch-4-is-now-an-fda-class-2-medical-device-detects-falls-irregular-heart-rhythm/#30ff9a2d2071
5. Ford Adding EksoVest Exoskeletons to 15 Automotive Plants: www.therobotreport.com/ford-eksovest-exoskeletons-automotive/
6. Ottobock reaches for growth with industrial exoskeletons: https://uk.reuters.com/article/us-ottobock-exoskeletons-focus/ottobock-reaches-for-growth-with-industrial-exoskeletons-idUKKCN1LR0LI
7. MIT's new robot takes orders from your muscles. *Popular Science*: www.popsci.com/mit-robot-senses-muscles/
8. This robotic tail gives humans key abilities that evolution took away: www.nbcnews.com/mach/science/robotic-tail-gives-humans-key-abilities-evolution-took-away-ncna1041431

9. Superhuman Vision: Bionic Lens. *Medium*: https://medium.com/@tinaphm7/superhuman-vision-bionic-lens-ad405fc42127

10. Samsung patents "smart" contact lenses that record video and let you control your phone just by blinking. *The Telegraph*: www.telegraph.co.uk/technology/2019/08/06/samsung-patents-smart-contact-lenses-record-video-let-control/

11. Florida Man Becomes First Person to Live With Advanced Mind-Controlled Robotic Arm: https://futurism.com/mind-controlled-robotic-arm-johnny-matheny

12. Robotic Hand Restores Wearer's Sense of Touch. *Smithsonian*: www.smithsonianmag.com/smart-news/robotic-hand-restores-wearers-sense-touch-180972737/

13. Artificial skin can sense 1000 times faster than human nerves. *New Scientist*: www.newscientist.com/article/2210293-artificial-skin-can-sense-1000-times-faster-than-human-nerves/

14. 7 human organs we can grow in the lab: https://blog.sciencemuseum.org.uk/7-human-organs-we-can-grow-in-the-lab/

15. 5 Most Promising 3D Printed Organs For Transplant: https://all3dp.com/2/5-most-promising-3d-printed-organs-for-transplant/

16. Facebook Takes First Steps in Creating Mind-Reading Technology: www.extremetech.com/extreme/296832-facebook-takes-first-steps-in-creating-mind-reading-technology

17. Elon Musk Announces Plans to "Merge" Human Brains With AI: www.vice.com/en_us/article/7xgnxd/elon-musk-announces-plan-to-merge-human-brains-with-ai

TREND 4
BIG DATA AND AUGMENTED ANALYTICS

The One-Sentence Definition

In very simple terms, "big data" refers to the exponential explosion in the amount of data being generated in this increasingly digital age, while "augmented analytics" refers to the ability to automatically work with and generate insights from data.

What Is Big Data and Augmented Analytics?

Let's start with the data itself, because data is critical to so many of the trends in this book, including artificial intelligence (AI, Trend 1), the Internet of Things (IoT, Trend 2), natural language processing (Trend 10), and facial recognition (Trend 12). Without data, the massive leaps we've seen in these trends – and many other technology trends – wouldn't be possible.

At the heart of big data is the idea that the more data you have, the easier it is to gain new insights, and even predict what will happen in the future. By analyzing masses of data, it's possible to spot patterns and relationships that were previously unknown. And when you can understand the relationships between data points, you can better predict future outcomes, and make smarter decisions on what to do next.

It's no exaggeration, then, to say that big data brings incredible opportunities to understand and change our world for the better.

But what is it that makes data, well, "big"? After all, data isn't exactly a new thing. What's new is the unprecedented digitization of our lives, where almost everything we do leaves a digital footprint. This is largely thanks to the rise of computers, smart phones, the internet, the IoT, sensors, and so on. Think of everyday activities like shopping online, reading the news in an app, paying for the morning coffee by card, messaging friends and family, taking and sharing photos, watching the latest show on Netflix, asking Siri a question, swiping right on a potential love match…we're all generating data all the time.

The sheer volume of data that we're creating, and the rate at which that volume is accelerating, is so vast that 90% of the data available in the world today was generated in the last two years.[1] What's more, every two years we're doubling the amount of data we have available.[2]

How much data are we talking about? Well, we're no longer talking about data in terms of gigabytes. These days, we're talking about terabytes (just over 1,000 gigabytes), petabytes (a little over 1,000 terabytes), exabytes (roughly 1,000 petabytes), and zettabytes (approximately 1,000 exabytes). According to market intelligence company IDC, the amount of data in the world could grow from 33 zettabytes in 2018 to 175 zettabytes in 2025.[3] To put that in perspective, if you stored 175 zettabytes on DVDs, you'd have a stack of DVDs so big it could encircle Earth 222 times! And the amount of data we're generating is likely to accelerate further. In other words, big data is only going to get bigger.

Our ever-increasing digital footprint has also given rise to another interesting aspect of big data: the fact that there are many new types of data that can be analyzed. We're no longer just working with numbers in spreadsheets, or entries in a database; today, "data" includes photo data, video data, conversation data (i.e. asking Alexa to play a certain song), activity data (such as browsing online or swiping left or right), and text data (like social media updates). Increasingly, the data we have to work with is *unstructured*, which means it can't be easily classified into neat rows and columns, like in a spreadsheet. This unstructured data is more challenging to analyze – which is a major problem when you consider that data is pretty much useless unless we can find a way to extract meaningful insights from it.

This is where the augmented analytics part comes in. Handling masses of data can be an expensive, time-consuming, and highly specialized task. In other words, there are some serious barriers between the data itself and the ability to turn that data into actionable insights. Augmented analytics is about breaking down those barriers and making it easier to generate amazing insights from data.

In a nutshell, augmented analytics involves using AI and machine learning (see Trend 1) to automate analytics processes, including gathering data from raw data sources, preparing and cleaning that data, building unbiased analytics models, and generating and communicating insights to those who need them. What's really exciting about this is it makes it easier for people to interact with data and extract the information they need, without the involvement of data specialists. So, in theory, with an augmented analytics tool, a non-tech expert could simply ask the system a question – like "Which of our employees are most likely to leave in the next 12 months?" – and the system would automatically generate a response.

Gartner predicts that by the end of this year, 40% of data science tasks will be automated,[4] meaning augmented analytics is on track to become the leading analytics method of the future. As the trend really takes off, it's likely we'll see many more specialized augmented analytics apps and tools designed for specific industries in the future. This is good news for businesses, since augmented analytics provides a way for organizations of all shapes and sizes to handle the vast amounts of complex data they're inundated with and give people in the organization easy access to analytics and insights from data. This wide access to data and insights is known as *data democratization*.

How Is Big Data and Augmented Analytics Used in Practice?

Now might be a good time to mention that, personally, I prefer the term "data" to "big data." The "big" implies it's the sheer volume of data that's really important. But equally important, if not more, is what we do with data. And, boy, can we do impressive things with data these days. Data, coupled with other trends like AI, is transforming our world – it's helping to making our homes smarter (see IoT, Trend 2), physically augment humans (see Trend 3), and build the smart cities of the future (see Trend 5), and that's just for starters. Data is also changing the way we do business.

Let's look at the main ways in which businesses can leverage data (big or otherwise) to their advantage.

Informing Business Decisions

Making better business decisions is absolutely one of the top priorities for most of the clients I work with. From how to hire the right people and target the right customers, to how to boost revenue, success means making the best decisions for your business. With data, you can better understand what's happening in the business and the wider market and predict what might happen in the future – information that's critical to good decision-making. Therefore, across every business function, data can and should be used to make smarter business decisions.

In one very simple example, US restaurant chain Arby's discovered that its renovated restaurants made more money than its unrenovated restaurants. Based on this knowledge, the company decided to carry out five times more restaurant remodels over the course of a year.[5]

Better Understanding Customers and Trends

The better you understand your customers, the better you can serve them. Sales and marketing activity is often based on past sales history – effectively, which customers previously bought which products or services. But, thanks to big data and augmented analytics, this activity is increasingly becoming more predictive. In other words, companies are now confidently and accurately anticipating what customers will want in the future. Netflix predicting what you might want to watch next is one simple example of this.

In another example, German retail company Otto discovered that customers are less likely to return items when they arrive within two days, and when they receive all their items at once, rather than in multiple shipments. Hardly earth shattering – keeping goods in stock and shipping efficiently makes good sense. However, Otto is like Amazon in that it sells products from many, many brands, which means stocking and shipping products all at once is a major challenge. So Otto analyzed the data from 3 billion past transactions, plus factors like weather data, to build a model that could predict what customers would want to buy in the next 30 days. Not only could the system do this, it could do so with 90% accuracy.[6] Now, the company can order the right products ahead of time and, as a result, product returns have been reduced by over 2 million items a year.

Delivering More Intelligent Products and Services

When you know more about your customers, you can give them exactly what they want: smarter products and services that respond intelligently to their needs. This has given rise to a wealth of smart products, such as smart speakers, smart watches, even smart lawnmowers. For plenty of examples of smart products and services in action, circle back to Trends 2 (IoT) and 3 (wearables), or turn to Trend 18 (digital platforms).

Improving Internal Operations

Every business process and every aspect of business operations can be streamlined and enhanced, thanks to big data. Optimizing pricing, accurately forecasting demand, reducing employee turnover, boosting productivity, strengthening the supply chain – across all areas of the business, it's easier than ever to make improvements, generate efficiencies, save money, automate processes, and more.

Remember the Otto example of predicting demand in order to improve stock ordering? Thanks to data (and more than a bit of AI, see Trend 1), this impressive process happens automatically. The company's system orders around 200,000 products a month without human intervention.

In another example, Bank of America worked with Humanyze (formerly Sociometric Solutions) to implement smart employee name badges, fitted with sensors that can detect social dynamics in the workplace. From the data generated, the bank noticed that top-performing employees at call centers were those who took breaks together. As a result, it instituted new group break policies and performance improved 23%.[7] You can find more examples of enhanced and automated business processes in Trend 13 (robots and cobots).

Creating Additional Revenue

Optimizing business processes, making better business decisions, and so on, will no doubt have a positive impact on the bottom line. But the link between data and the bottom line can be much more explicit, meaning data can be monetized to create new revenue streams.

This may encompass bringing new data-driven products to market (such as the smart products outlined in Trends 2 and 3), or it could mean actively selling data through optimized services (such as

Google's data-driven advertising offering). Data can even increase the value of a company; at the time of writing, the world's top three most valuable brands were Google, Apple, and Amazon – each of them data-driven businesses.[8]

Key Challenges

You might think that some of the most obvious challenges around big data are the technology, infrastructure, and skills challenges. To put it another way, do you have to have the budget, infrastructure, and know-how of, say, Google or Amazon to benefit from big data? Thanks to augmented analytics and big-data-as-a-service (BDaaS), the answer is no. I've covered augmented analytics earlier in the chapter, so let's briefly look at BDaaS. The term refers to the delivery of big data tools and technology – and potentially even data itself – through software-as-a-service platforms. These services allow companies to access big data tools without the need for expensive infrastructure investments (see also AI-as-a-service in Trend 1), thereby helping to make big data accessible to even small businesses. This also helps to overcome the massive skills gap in big data. Essentially, there aren't enough data scientists to go around; the McKinsey Global Institute predicts that, by 2024, there'll be a shortage of approximately 250,000 data scientists – and that's just in the US.[9]

As analytics tools advance, my hope is that technology, infrastructure, and skills will become less daunting barriers to working with data. But that doesn't mean there won't be other barriers to contend with. I believe two of the biggest challenges around big data are data security and privacy.

Ever-growing volumes of data – and the fact that data is becoming more of a critical business asset – brings huge challenges in terms of protecting that data. It's therefore vital that organizations take steps to protect their data from attack, particularly when it comes to personal data (like customer or employee data). Advances like the IoT add an extra dimension to the threat, since many connected devices are totally unsecured, thereby providing a potential way in for hackers. (One study has found that 82% of organizations believe that unsecured IoT devices will cause a data breach in the next few years.[10]) But your employees are another significant threat to consider. So as well as having a robust data security policy in place, it's vital you raise awareness of the potential threats and educate your teams on the need to protect data.

Security is closely linked to data privacy, since so much of the data that organizations are working with contains personally identifiable information. Regulators are, to some extent, still playing catchup when it comes to data privacy laws, but that will change. Recent GDPR guidance in Europe is designed to promote the safe and ethical handling of personal data – and give individuals a greater say in how organizations use their data. Therefore, it's not enough to protect your data securely – you also need to take an ethical approach to collecting and using that data. This means being completely transparent, making customers and other stakeholders aware of what data you're gathering and why, and giving them the chance to opt out where possible. Those companies who don't comply with tightening regulation, or who play fast and loose with people's data, risk serious financial and reputational blowback in the future.

How to Prepare for This Trend

Despite the challenges, most experts, myself included, believe the benefits of big data are huge. Data can bring enormous value to your organization, providing you prepare properly. For me, this means:

- Improving data literacy across the organization

- Creating a data strategy

Let's look at each step in turn.

Improving Data Literacy Across the Organization

The more data literate your organization is, the better your results will be. It's as simple as that. But that doesn't mean everyone has to be a data scientist. It simply means that everyone right across the business must be comfortable with data: talking about data, using data, thinking critically about data, pulling meaningful insights from data, and ultimately acting on what data tells them. Data literacy is about everyone putting data to use, essentially.

Raising data literacy across the business is a case of establishing your current levels of data literacy, communicating why data literacy is important, identifying data advocates who can sing the praises of data, ensuring access to data, and educating those across the business on how to get the most out of data.

Creating a Data Strategy

It's also vital you have a data strategy in place. A data strategy helps you remain focused on the data that matters most to your business – as opposed to collecting data on anything and everything, which is rather an expensive way to go about it! With so much data available these days, the trick is to focus on finding the exact, specific pieces of data that will best benefit your organization. A data strategy helps you do just that. With a robust data strategy you can set out how you want to use data in practice, clarify your top data priorities, and chart a clear course to achieving your goals.

Your data strategy must be unique to your business, but, broadly speaking, I'd expect a good data strategy to cover the following points:

- **Business needs.** To truly add value, data must be driven by specific business needs, which means your data strategy must be driven by your overarching business strategy. Basically, what is your business trying to achieve, and how can data help you achieve those strategic objectives? Here, it's wise to identify no more than three to five key ways in which data can help the business achieve its strategic goals, answer key business questions, or overcome its main challenges. Then, for each data use, you then identify the following…

- **Data requirements.** What data do you need to achieve your goals and where will that data come from? Do you, for example, already have the data you need? Do you need to supplement internal company data with externally available data (such as social media data)? If you need to collect new data, how will you go about that?

- **Data governance.** This is what stops your data becoming a serious liability, and involves considerations such as data quality, data security, privacy, ethics, and transparency. For example, who is responsible for making sure your data is accurate, complete, and up to date? What permissions do you need to secure in order to gather and use the data?

- **Technology requirements.** In very simple terms, this means looking at your hardware and software needs for collecting data, storing and organizing data, analyzing data, and communicating insights from data.

- **Skills and capacity.** Do you have the skills to deliver your data needs and, if not, how will you overcome the skills gap? Will you, for example, need to hire new people, or can you partner with external data providers?

Notes

1. How Much Data Does The World Generate Every Minute? *IFL Science*: www.iflscience.com/technology/how-much-data-does-the-world-generate-every-minute/
2. The future of big data: 5 predictions from experts: www.itransition.com/blog/the-future-of-big-data-5-predictions-from-experts

3. Data Age 2025: The Digitization of the World, IDC: www.seagate.com/files/www-content/our-story/trends/files/idc-seagate-dataage-whitepaper.pdf

4. Gartner Says More Than 40 Percent of Data Science Tasks Will be Automated by 2020: www.gartner.com/en/newsroom/press-releases/2017-01-16-gartner-says-more-than-40-percent-of-data-science-tasks-will-be-automated-by-2020

5. Arby's forecasts retail success in Tableau, leading to 5x more renovations in a year: www.tableau.com/solutions/customer/renovating-retail-success-arbys-restaurant-group

6. German ecommerce company Otto uses AI to reduce returns: https://ecommercenews.eu/german-ecommerce-company-otto-uses-ai-reduce-returns/

7. The Quantified Workplace: Big Data or Big Brother? *Forbes*: www.forbes.com/sites/bernardmarr/2015/05/11/the-nanny-state-meets-the-quantified-workplace/#5b16648669fa

8. Amazon beats Apple and Google to become the world's most valuable brand: www.cnbc.com/2019/06/11/amazon-beats-apple-and-google-to-become-the-worlds-most-valuable-brand.html

9. The age of analytics: Competing in a data-driven world: www.mckinsey.com/business-functions/mckinsey-analytics/our-insights/the-age-of-analytics-competing-in-a-data-driven-world

10. 2018 study on global megatrends in cybersecurity: www.raytheon.com/sites/default/files/2018-02/2018_Global_Cyber_Megatrends.pdf

TREND 5
INTELLIGENT SPACES AND SMART PLACES

The One-Sentence Definition

Intelligent spaces and smart places are physical spaces – such as homes, office buildings, or even cities – that have been kitted out with technology to create an intelligent, connected environment.

What Are Intelligent Spaces and Smart Places?

There's no doubt that our physical spaces are getting smarter. Our homes are now stocked with intelligent devices, such as smart speakers, that learn our behavior and preferences and react accordingly. But the trend for intelligent spaces extends far beyond our homes. Workplaces are becoming smarter. Entire buildings are being transformed into connected spaces that respond intelligently to the people who live and work inside. Even cities are becoming smart, thanks to initiatives like smart street lighting and intelligent traffic networks.

As a trend, intelligent spaces and smart places are inextricably linked with other trends in this book: artificial intelligence (AI, Trend 1), the Internet of Things (IoT, Trend 2), automation (Trend 13), autonomous vehicles (Trend 14), and advanced connectivity and 5 G

(Trend 15). It's the combination and acceleration of these advances that allows us to create spaces in which humans and technology interact in a more intelligent, connected, and automated way.

What does this mean in practice? A simple example is smart office lighting that comes on when workers are in the vicinity and turns off automatically when there's no one around. A more complex example might be an airport that's fully decked out with the latest connected technology, with automated kiosks for self-check-in, self-service areas to drop off luggage, a facial recognition system to improve and automate some aspects of security, AIs that automatically track the flow of people and monitor queue length, and a dedicated passenger app that provides updates on flights and in-airport services. There are lots of real-life examples coming up later in the chapter – from the relatively simple to advanced, citywide initiatives.

These days, pretty much any space can be made more connected and intelligent. Offices, factories, hotels, hospitals, transport hubs, apartment buildings, individual homes, shopping centers, schools, libraries…you name it. But why would we want to make all these spaces intelligent? The benefits of smart places include increased energy efficiency, increased productivity, greater quality of life, increased safety, and simplified processes. Generally, the idea is to make everyday life easier and better for the people who use these spaces (be they residents, commuters, employees, clients, or whomever), while maximizing efficiency and the use of resources.

It's worth noting that the definition of a smart space is occasionally extended to include digital environments or platforms that create a collaborative digital experience not confined to a single computer or device – for example, an online platform that allows colleagues to communicate with each other and share content seamlessly. However, for the purposes of this chapter, I'm focusing on physical spaces that have been enhanced with technology.

How Are Intelligent Spaces and Smart Places Used in Practice?

An intelligent space could be anything from a small apartment to an entire building or even a whole city. Let's take a look at how our physical environments are becoming smarter.

Smart Homes

The range of smart home gadgets is constantly growing. Just about every household appliance is now available in a connected, smart version, from the fairly commonplace smart thermostat to less obvious offerings, such as smart washing machines, smart lawnmowers, and even smart toilets. Circle back to Chapter 2 to read more about the IoT-enabled devices that are finding their way into our homes.

Smart Offices and Buildings

By incorporating smart technology into offices, workplaces, and other buildings, we can enhance the local environment, change the way people interact with buildings, and create a better user experience – thereby improving productivity, safety, and well-being in the process. Smart building technology takes many forms and may be used to automate certain processes – for example, security – or to enhance human decision-making, often in real time. Let's look at some of the smart systems that can be incorporated into buildings today:

- IoT sensors can be used to detect and monitor occupancy, and gain insights on building usage. In fact, sensors underpin many of the examples in this chapter, as they enable innovations such as **smart lighting** – a key part of any smart building scheme. According to Gartner, smart lighting has the potential to reduce energy costs by 90%.[1] These smart lighting systems can be used to turn the lights on or off according to motion in the room, and also to automatically adjust lighting levels according to daylight levels.

- **Smart climate control systems** can regulate the temperature of buildings automatically according to, among other factors, occupants' usage patterns.

- **Smart desks** can perform a range of functions, such as adapting to the user's preferences when they "log in" to their desk (e.g. a taller person's desk may automatically raise itself up). They can also gather data on the amount of time users spend sitting or standing, and prompt users with an alert if they've been sitting too long. And for offices that have adopted hot-desking or desk sharing, some smart desks come with an online booking platform that allows users to search for and reserve vacant desks.

- If you've got a smart desk, why not add a **smart chair**? Sensors located in intelligent office chairs can monitor the user's posture and provide feedback to help improve posture and reduce the risk of back pain.

- From office blocks and warehousing facilities to apartment blocks, security is a key consideration for most modern buildings. Today, **smart locks** can eliminate the need for people to carry keys or key cards (which can easily be lost or stolen). And advanced **facial recognition systems** (see Trend 12) can be used to automatically identify who is and isn't allowed access to the building, and keep track of visitors – thereby eliminating any need for ID badges.

- Some employers are also getting in on the **wearable technology** trend (Trend 3), and offering their employees free or discounted fitness trackers in order to encourage healthier lifestyles.

Let's look at two brief examples of how these sorts of technologies have been successfully deployed in real life:

- **Microsoft's** renovation of its Amsterdam headquarters shows how smart office technology can deliver tangible benefits. Prior to the renovation project, sensors were used to monitor how desks, meeting rooms, and communal areas were used by occupants, giving Microsoft incredible insight into how people actually used the office. Thanks to this data, Microsoft was able to reduce the amount of space it needed, freeing up one-and-a-half floors to let to another company.[2] Now that the building has been renovated, sensors are still used to monitor occupancy, as well as temperature, noise levels, light levels, and more.

- Dubai's **Burj Khalifa** – at the time of writing, the world's tallest building – incorporates several smart building technologies. For example, the building's automation system delivers real-time data to an analytics platform that analyzes the data for potential maintenance issues. Thanks to this system, facility managers have been able to improve building maintenance while reducing total maintenance hours by 40%.[3]

Smart Cities and Smart City Initiatives

A city that leverages technology to increase efficiencies and improve the quality of services and life for its residents? That's a smart city. Just like our homes and businesses, our cities can use the masses of data generated and powerful technologies like AI to gain actionable insights on how to save time, money, and energy.

More and more of us are now living in cities – the UN predicts that 68% of the world's population will live in urban areas by 2050[4] – which means our cities are facing growing environmental, societal, and economic challenges. Smart city initiatives offer a way to overcome these challenges. In fact, a report by McKinsey Global Institute found cities can use smart technologies to improve key quality of life indicators – such as the daily commute, health issues, or crime incidents – by 10–30%.[5] It's no wonder that more cities are embracing smart technology; one survey by the National League of Cities found that 66% of cities have invested in smart city technology in some capacity.[6]

This technology is commonly retrofitted, but as new cities are built, the technology can be incorporated into cities from the very start.

But what do we mean by smart city initiatives? These initiatives can cover anything from power distribution, transport systems, and even rubbish collection. Here are a few inspiring examples from around the world:

- Traffic is the bane of many a city-dweller's life, but technology offers some promising solutions. For example, public transportation routes can be adjusted in real time according to demand, or traffic flows can be monitored and analyzed to improve congestion via intelligent traffic light systems. **Alibaba's City Brain** system uses AI to optimize a city's infrastructure, and in the Chinese city of Hangzhou, it has helped reduce traffic jams by 15%.[7]

- Smart streetlamps can illuminate the area when sensors detect cars or pedestrians and dim when no one is around. The intelligent streetlamps developed by **GE** are just one example of smart street lighting.[8]

- In Denmark's **Middelfart Municipality**, the city is collecting energy efficiency data from city properties, including information on the indoor climate, energy usage, and maintenance of the buildings. The data is used to make adjustments for better energy efficiency.[9]

- Mobile and broadband company **Telefonica** has invested heavily in the smart city concept in its home country, Spain. In one example, sensors are attached to refuse containers to report, in real time, how full they are – which allows city officials to allocate waste collection resources more efficiently. Locals can also tag overflowing rubbish containers in their neighborhood using an app. And in Valencia, Telefonica is helping to tackle the city's parking issues; sensors in parking spaces help monitor capacity and give officials real-time data on the density of parking across the city.

- The **Amsterdam Smart City initiative**, which began in 2009, spans over 170 projects designed to improve the city's real-time decision-making.[10] Streetlamps have been upgraded so that lights can be dimmed according to pedestrian usage. Traffic sensors allow the city to inform drivers of current traffic conditions. And a number of homes have been given smart energy meters.

- My hometown, **Milton Keynes**, is working with more than 40 partners on smart city initiatives, including monitoring traffic and footfall through public spaces in order to plan public transport routes, footpaths, and cycle paths.[11]

- **Sidewalk Labs** – a smart city startup from Google's parent company Alphabet – is dedicated to improving urban infrastructure through technology. One of its latest plans is to turn a large section of Toronto's Lake Ontario shoreline into a highly efficient, innovative district, featuring publicly available wi-fi, heated pavements to automatically melt snow, self-driving delivery robots, and sensors that collect data on energy consumption, building use, and traffic patterns.[12] A highly connected, self-regulating neighborhood, in other words.

Key Challenges

For smart cities, adopting these new technologies is disruptive and expensive, and requires careful thought around regulations regarding use of the technology. For individual businesses, adoption is certainly easier – although still not without challenges.

As the smart home market is generally more established than connected workplaces, we can look to our homes for an indication of some of the key challenges businesses would need to overcome:

- **Wi-fi connectivity issues.** Smart devices in the home tend to rely on wi-fi connectivity in order to gather and transmit data and to link up with other devices. No internet, no Alexa, in other words. But as the notion of smart spaces expands to cover workplaces and public spaces, advances like edge computing (Trend 7) and 5 G networks (Trend 15) will help to overcome this issue.

- **Incompatibility between devices.** Key to the idea of an intelligent space is the ability for different devices and systems to seamlessly connect with each other and work together to create the ideal environment. But with so many providers piling into the market, achieving compatibility across the board may be a challenge.

- **Data security concerns.** Smart, connected spaces need data to function – data concerning where people are, what they're doing, and so on. As with any data that relates directly to individuals, it's vital this data is properly protected.

- **Privacy concerns.** Many consumers with Alexa devices were outraged in 2019 when it emerged contractors are paid to listen to (anonymized) recordings of people's Alexa interactions. Yes, we all know deep down that installing a smart speaker effectively means installing a recording device in your home, but the reality of a human being somewhere listening to what you've said made more than a few people uncomfortable.

That last issue, privacy, is a big one. As we're increasingly surrounding ourselves with devices that can track our activities and conversations, the implications for individual privacy are being more keenly felt. We all have a right to a certain level of privacy, whether we're in our own home, at our place of work, or even on a busy street. Campaigners argue that smart places infringe this right to privacy – particularly when there's a lack of transparency or individuals have no ability to opt in or out.

As an example, Sidewalks Labs' project to develop the high-tech neighborhood in Toronto has come up against stiff opposition, and a #Blocksidewalk campaign was launched by local citizens in 2019 with the goal of stopping the project altogether.[13] The campaigners argue that residents should be given a greater say in the deployment of such technology in cities, and that these decision-making processes should be transparent, rather than city leaders and technology companies doing deals behind closed doors. A big focus of locals' concerns is around privacy for residents and those passing through the area, plus the security of the data being captured.

And finally, businesses will also have to overcome the data and AI skills gap if they're to use these connected systems most effectively. People have to be trained to develop and use these systems properly – and the company as a whole must develop a mindset and culture that values this technology and recognizes the benefits it brings to everyone in the workplace.

How to Prepare for This Trend

Creating a smart, connected workplace can deliver serious benefits for business, such as increased productivity, improved efficiency, reduced operating costs, and increased employee satisfaction.

There's no one-size-fits-all approach to creating an intelligent workplace, as every business has different needs and its own unique environment. But the following tips may help you weigh up the possibilities for your business:

- **Look for success stories in your industry and beyond.** How have businesses like yours realized their vision of a smart workplace?

- **Consider your overarching business strategy.** What is your business trying to achieve and could smart technology help you get there?

- **Take it one step at a time.** Intelligent spaces require investments in infrastructure, so you'll likely need to prioritize certain areas of the business rather than trying to implement widespread solutions right across the business. Focus on your priority areas – which will be informed by your most pressing business needs – and then build from there.

Notes

1. Gartner Says Smart Lighting Has the Potential to Reduce Energy Costs by 90 Percent: www.gartner.com/en/newsroom/press-releases/2015-07-15-gartner-says-smart-lighting-has-the-potential-to-reduce-energy-costs-by-90-percent

2. Could a smart office building transform your workplace? *Raconteur*: www.raconteur.net/technology/smart-buildings-office-productivity

3. Smart Buildings: The Ultimate Guide: https://blog.temboo.com/ultimate-smart-building-guide/

4. 68% of the world population predicted to live in urban areas by 2050: www.un.org/development/desa/en/news/population/2018-revision-of-world-urbanization-prospects.html

5. Smart cities: Digital solutions for a more liveable future: www.mckinsey.com/~/media/mckinsey/industries/capital%20projects%20and%20infrastructure/our%20insights/smart%20cities%20digital%20solutions%20for%20a%20more%20livable%20future/mgi-smart-cities-full-report.ashx

6. Cities and Innovation Economy: Perceptions of Local Leaders: www.nlc.org/resource/cities-and-innovation-economy-perceptions-of-local-leaders

7. In China, Alibaba's data-hungry AI is controlling (and watching) cities, *Wired*: www.wired.co.uk/article/alibaba-city-brain-artificial-intelligence-china-kuala-lumpur

8. How smart is your street light?: www.ge.com/reports/25-06-2015how-smart-is-your-street-light/

9. 10 examples of smart city solutions: https://stateofgreen.com/en/partners/state-of-green/news/10-examples-of-smart-city-solutions/

10. 8 Years On, Amsterdam is Still Leading the Way as a Smart City, *Medium*: https://towardsdatascience.com/8-years-on-amsterdam-is-still-leading-the-way-as-a-smart-city-79bd91c7ac13

11. Milton Keynes: Using Big Data to make our cities smarter: www.bernardmarr.com/default.asp?contentID=728
12. A Big Master Plan for Google's Growing Smart City: www.citylab.com/solutions/2019/06/alphabet-sidewalk-labs-toronto-quayside-smart-city-google/592453/
13. Newly formed citizens group aims to block Sidewalk Labs project, *The Star*: www.thestar.com/news/gta/2019/02/25/newly-formed-citizens-group-aims-to-block-sidewalk-labs-project.html

TREND 6
BLOCKCHAINS AND DISTRIBUTED LEDGERS

The One-Sentence Definition

A blockchain or distributed ledger is, in very simplistic terms, a kind of highly secure database – a way of storing information, in other words.

What Are Blockchains and Distributed Ledgers?

IBM CEO Ginni Rometty has said, "What the Internet did for communications, I think blockchain will do for trusted transactions."[1] That's quite a strong prediction. So what's so special about blockchain technology?

In today's digital age, storing, authenticating, and protecting data presents serious challenges for many organizations. Blockchain technology promises a practical solution to this problem, providing a useful and secure way to authenticate information, identities, transactions, and more. As we'll see later in the chapter, this makes blockchain an increasingly attractive tool for industries like banking and insurance, among others. In fact, blockchain can be used to provide a super-secure real-time record of pretty much anything:

financial transactions, contracts, supply chain information, even physical assets.

Blockchain, then, is essentially a way of storing data. To put it in more technical terms, it's a form of open, distributed ledger (i.e. a database), where the data is distributed (i.e. duplicated) across many computers and is typically *decentralized*. It's this decentralization aspect that makes blockchain so transformative. For one thing, it means there's no one central point of attack for hackers to target – which is part of what makes blockchain super-secure. (While nothing is totally "hack-proof," blockchain represents a significant leap forward in information security.) But the decentralized nature of blockchain also means data can be verified by user consensus, in a peer-to-peer system, rather than being processed and controlled by one central administrator – more on this coming up.

In case you're wondering where the name comes from, records in a blockchain are called "blocks" and each block is linked to the previous block, forming the "chain." Every block has a time and date stamp, noting when the record was created or updated. The chain itself can be public (like Bitcoin) or private (like a banking blockchain) – this is a key point that I'll revisit later. And when changes are made to a block, the whole blockchain is kept in synch and each user's copy of the blockchain is updated in real time.

Whether the chain is public or private, users can only edit parts of the blockchain by possessing the cryptography key needed to alter the file. To illustrate how this works in practice, I often use medical records as an example. Imagine a digital medical record as a blockchain. Each entry (e.g. a diagnosis and treatment plan) is a separate block, with a time and date stamp that marks when the record was created, and only people with the cryptography key can access the information in that block. So, in this case, the patient might have the key that allows them to give a consultant and the patient's GP access to the information. Information can be shared with another party – say, between the consultant and the GP – but only with permission, using the secure key.

I briefly mentioned Bitcoin as an example of a public blockchain. Many people think blockchain and Bitcoin are the same thing, but they're not. The digital currency Bitcoin functions using blockchain technology – blockchain provides the public ledger for Bitcoin transactions – and Bitcoin was the first example of blockchain in use. But blockchain has many applications beyond cryptocurrency.

Before we delve deeper into practical uses of blockchain, there's another important distinction we need to make: the distinction between blockchains and *distributed ledger technology*. Strictly speaking, the two terms aren't quite interchangeable. Blockchains and distributed ledgers certainly overlap a great deal, which is why I've combined them into one chapter. Both, for instance, refer to information being distributed across a network, and both serve to enhance security. But there is a difference. It's perhaps more accurate to say that blockchain is one way to implement distributed ledger technology, but not the only way.

Here's the key difference: blockchains are generally public, which means anyone can participate in the chain and anyone can validate information – it's a truly decentralized, democratic system with no one body or person being "in charge." Bitcoin is the perfect example of this kind of public blockchain. With Bitcoin, transactions aren't verified by a trusted organization, like Visa or Mastercard, but by the Bitcoin community in a peer-to-peer system. A distributed ledger, on the other hand, could be private, meaning access is restricted by one centralized body (say, a company or government agency). So a distributed ledger isn't necessarily decentralized and democratic – but the information is still distributed and generally far more secure than in traditional databases. A good way to sum up the difference is: a blockchain is typically open and *permissionless,* while a distributed ledger tends to be *permissioned.*

For ease, I use the term "blockchain" throughout this chapter. However, strictly speaking, many of the applications and examples outlined in this chapter are examples of private distributed ledgers, rather than a public blockchain. Either way, the technology promises to revolutionize many aspects of how we do business. Blockchain fans predict it could be as disruptive as the internet was before it.

How Are Blockchains and Distributed Ledgers Used in Practice?

While it's true that early adopters of blockchains have been using the technology for financial transactions, we're likely to see blockchain being used far more widely in the next few years. Medical records, transfer of ownership records, property transactions, HR records – any process of recording, overseeing, and verifying information could be enhanced by blockchain technology. In theory, any centralized, unwieldy, and unsecure ledger system could be replaced with a streamlined, distributed blockchain system for record keeping. What's more, Blockchain's approach to encrypting information could also be adopted as a way to secure IoT devices (see Trend 2).

Let's explore some real-world examples of how organizations are beginning to use blockchain technology to their advantage.

Improving Insurance through Smart Contracts

We already know from Bitcoin that blockchain is great at facilitating transactions, but it can also be used to formalize commercial relationships through smart contracts that automatically execute when agreed-on conditions are filled. Smart contracts have the potential to revolutionize the insurance industry by ensuring only valid claims are paid out; for example, with claims and policies stored on a blockchain, the blockchain would know instantly whether multiple claims have been filed for the same accident. Then when conditions for a valid claim have been met, the claim could be paid out automatically, without any human intervention.

Here are some examples of insurers embracing blockchain:

- **Nationwide** insurance company is trialing a proof-of-insurance blockchain solution called RiskBlock that would allow law enforcement and other insurers to verify insurance coverage in real time.[2]

- Insurer **AIG** is working with IBM to pilot a blockchain-based smart contract system for multinational insurance policies.[3] Multinational insurance can be complex because of the different jurisdictions involved, but AIG is confident the new system will enable real-time sharing and updating of information between the main policyholder and overseas subsidiaries.

- Shipping and transport consortium **Maersk** has unveiled a blockchain solution for streamlining marine insurance.[4]

Protecting Ownership, Including Intellectual Property

There are many potential uses for blockchain in verifying ownership of assets – and even in transferring ownership of those assets.

- **Kodak** appears to be rebranding itself as a blockchain business. The company has announced a platform that will record and track usage of photographs across the internet, allowing rights holders to seek payment when their work is used without permission.[5]

- **Mycelia** is a blockchain-based solution designed to record and track royalties for musicians, and allow musicians to create a record of ownership for their work.[6]

Verifying People's Identities and Credentials

It's not just assets that can be verified using blockchain; identities and other personal information can be securely stored and verified.

- **APPII** has launched what it calls the "world's first blockchain career verification platform," designed to cut the amount of time employers spend checking candidates' qualifications and experience.[7] Candidates create a profile, list their education, professional experience, accreditation, and so on. Then, previous employers and educational institutions can verify the information, meaning the company hiring that candidate doesn't need to check all the details again. The platform also uses facial recognition technology to verify a candidate's identity. Will all this eliminate the risk of hiring people that have, shall we say, "embellished" their CV? Time will tell.

- **Sierra Leone's government** has announced it is adopting a blockchain-based national identity tool, called the National Digital Identity System.[8] In the first phase, citizens' identities will be digitized, then the system will be used to create universally recognized national identification numbers that can't be duplicated or reused. The hope is this system will unlock access to credit and financial services for citizens.

Promoting Traceability in the Supply Chain

Blockchain can bring transparency to the supply chain and provide a complete record of the life cycle of products, so it's no wonder the technology has been enthusiastically welcomed by industries and companies that want to demonstrate the provenance of their goods.

- The **Everledger Diamond Time-Lapse Protocol** tracks a diamond from the mine to the store, and has already been used to record the details of more than 2 million diamonds.[9]

- **Walmart** is using blockchains to track the safety of leafy greens. Farmers input detailed records of their produce into a blockchain and, in the event of a future contamination (like when *E. coli* was found in Walmart lettuce in 2018), Walmart will be able to pinpoint affected batches much more easily.[10]

- **Blockverify** is a blockchain solution designed to bring transparency to the supply chain and identify counterfeit and stolen goods. So far, the platform has found uses in the diamond, pharmaceuticals, electronics, and luxury goods markets.[11]

- Sustainable shoe brand **CANO** is using Oracle's blockchain platform to provide transparency around its supply chain, recording every step of the shoe-making process, from what materials are involved to who's making the shoes.[12]

- With **VeChain's Wine Traceability Platform**, wine bottles can be fitted with an encrypted chip that contains the product's information on a blockchain, with the info being verified by third-party auditors. Australian winemaker Penfolds is one of the winemakers on board with the initiative.

Improving the Banking Industry

With blockchain's reputation for making secure transactions easy, it makes sense that the banking industry is exploring many blockchain uses.

- **Barclays** has launched a number of blockchain initiatives for tracking financial transactions, compliance, and fraud. The company is so convinced of the merits of blockchain, it's described the technology as a "new operating system for the planet."[13]

- **Bank Hapoalim**, one of Israel's largest banks, has been working with Microsoft to create a blockchain for managing bank guarantees for large purchases, like property.[14]

- **Shinhan Bank**, South Korea's oldest bank, is developing a blockchain-based personal stock lending service.[15]

Cutting out the Middleman

If I was an aggregator company like Uber, Airbnb, or Expedia, I'd be very worried about the impact of blockchain technology on my business model, because blockchains can be used to create a secure, decentralized way for service providers and customers to connect directly and transact in a safe environment – without the need for a middleman like, say, Uber.

- **TUI Group** has confirmed blockchain will be a key part of its business model going forward, eventually eliminating the need for intermediaries like Expedia. One pilot project, BedSwap, allows hotels to record hotel inventories and availability in real time on a blockchain.[16]

- **OpenBazaar** is a decentralized marketplace where goods and services can be traded with no middleman, using cryptocurrency.[17]

- Hotel aggregator **GOeureka** is using blockchain to increase transparency and cut costs – by giving users access to 400,000 hotel rooms with no middleman commission costs.[18] It'll be interesting to see if other intermediary-type businesses leverage blockchain in a similar way to protect their business model.

Making Cryptocurrency More Mainstream

We know that blockchain is the technology that underpins Bitcoin, but as the wider uses of blockchain are becoming more apparent, the cryptocurrency arena is also growing.

- Boxer **Manny Pacquiao** has unveiled his own cryptocurrency, called Pac, which will allow fans to buy Manny merchandise and engage with their hero.[19]

- In 2019, Facebook announced it was venturing into virtual currency with **Libra**.[20] The idea is that Libra will become a global currency – virtual money that you can transfer to people or use to buy things – potentially benefiting those in developing countries where access to banks and financial services is limited. Yet, unlike Bitcoin, Libra won't be decentralized, not at first anyway. It'll be controlled by Facebook and its partners, which has prompted some concerns that Facebook could use transaction data to its own benefit (such as ad targeting). Regulators have also been piling on the pressure, prompting some of Libra's backers to pull out and throwing the whole project potentially into crisis.[21]

Key Challenges

There will likely be challenges around industry regulation – just look at the scrutiny Facebook has attracted over its Libra cryptocurrency plans – and we can expect regulators to pay greater attention to blockchains in future.

But, right now, the biggest challenge for those looking to adopt the technology is the fact that blockchain is still in its infancy. In fact, it's fair to say the technology is about as mature as the internet was in 1996. We're a long way off blockchains becoming the norm, in other words. Those who rush in at this stage, without a clear plan of how they want to use blockchain technology and what they want to achieve, could end up wasting a lot of time and money.

It's the old "fear of missing out" thing. Companies who keep hearing blockchain is the next big thing become desperate to demonstrate how cutting edge they are. As a result, they dive in and adopt new practices that are badly conceived and ultimately don't deliver real value. So, while I firmly believe blockchain has the potential to transform many aspects of business, that transformation is likely to be gradual. There will undoubtedly be false starts and failures along the way.

How to Prepare for This Trend

In the long run, I believe blockchain will bring many benefits to businesses, including:

- Reduced costs – by reducing or eliminating the need for "middleman" services, blockchain promises to remove the financial burdens involved in making and recording transactions.

- Increased traceability – because every point of a supply chain could in theory be reliably recorded in a blockchain.

- Enhanced security – thanks to blockchain's encryption, handling and securing sensitive data is likely to become a lot easier.

Although it may take years for blockchains to become commonplace, companies can't afford to be caught out by what's coming. When it fully takes off, the impact will be significant – just as it was with the internet. Therefore, I'd advise all business leaders to keep abreast of what's happening in blockchain technology and begin to consider the practical implications of blockchain for their business.

Notes

1. @IBM, Twitter: https://twitter.com/ibm/status/877599373768630273?lang=en
2. Nationwide delves into blockchain with consortium partners: www.ledgerinsights.com/nationwide-insurance-blockchain-consortium-riskblock/
3. AIG teams with IBM to use blockchain for "smart" insurance policy: https://www.reuters.com/article/aig-blockchain-insurance/aig-teams-with-ibm-to-use-blockchain-for-smart-insurance-policy-idUSL1N1JB2IS

4. World's first blockchain platform for marine insurance: www.ey.com/en_gl/news/2018/05/world-s-first-blockchain-platform-for-marine-insurance-now-in-co

5. KOKAKOne: https://www.kodakone.com/

6. Mycelia: Imogen Heap's Blockchain Project for Artists & Musicians: http://myceliaformusic.org/2018/06/20/mycelia-imogen-heaps-blockchain-project-artists-music-rights/

7. Blockchain-Based CVs Could Change Employment Forever: https://bernardmarr.com/default.asp?contentID=1205

8. Sierra Leone Aims to Finish National Blockchain ID System in Late 2019: https://cointelegraph.com/news/sierra-leone-aims-to-finish-national-blockchain-id-system-in-late-2019

9. Diamond Time-Lapse Protocol, Everledger; https://www.everledger.io/pdfs/Press-Release-Everledger-Announces-the-Industry-Diamond-Time-Lapse-Protocol.pdf

10. In Wake of Romaine E. coli Scare, Walmart Deploys Blockchain to Track Leafy Greens: https://corporate.walmart.com/newsroom/2018/09/24/in-wake-of-romaine-e-coli-scare-walmart-deploys-blockchain-to-track-leafy-greens

11. Blockverify: http://www.blockverify.io/

12. Oracle Blockchain Platform Helps Big Businesses Incorporate Blockchain, *Forbes*: www.forbes.com/sites/benjaminpirus/2019/07/22/oracle-blockchain-platform-helps-big-businesses-incorporate-blockchain/#4dfd6668797b

13. Why blockchain could be a new "operating system for the planet": https://home.barclays/news/2017/02/blockchain-could-be-new-operating-system-for-the-planet/

14. Simplifying blockchain app development with Azure Blockchain Workbench: https://azure.microsoft.com/en-gb/blog/simplifying-blockchain-app-development-with-azure-blockchain-workbench-2/

15. South Korea's Shinhan Bank Developing a Blockchain Stock Lending System: https://cointelegraph.com/news/south-koreas-shinhan-bank-developing-a-blockchain-stock-lending-system

16. TUI Utilizes Blockchain Technology To Reshape The Travel Industry, *Medium*: https://medium.com/crypto-browser/tui-utilizes-blockchain-technology-to-reshape-the-travel-industry-fb83ba5395bf

17. OpenBazaar: https://openbazaar.com/

18. GOeureka uses blockchain to unlock 400,000 hotel rooms with zero commission: https://venturebeat.com/2018/09/28/goeureka-uses-blockchain-to-unlock-400000-hotel-rooms-with-zero-commission/

19. Boxer Manny Pacquiao intros cryptocurrency to cash in on his fame: www.engadget.com/2019/09/01/boxer-manny-pacquiao-cryptocurrency/?guce_referrer=aHR0cHM6Ly93d3cuZ29vZ2xlLm NvbS8&guce_referrer_sig=AQAAAK7vAztV_YQ8CCRaNRabPRV0 w4v6NKkDsm1TNN1S_6uft7QnpDAP4q8djMIiT0UddbThlhR60uTs VAfUwpWEKtZ4zN9abhux_HiHq2jfOvYt3UVQasGkGKJ247jzJOhr PseNvjZ2rEnPlD_ARgvYKDnTD1CQ0KSTO0Al8l9lgMpK& guccounter=2

20. What is Libra? Facebook's cryptocurrency, explained, *Wired*: www.wired.co.uk/article/facebook-libra-cryptocurrency-explained

21. Where it all went wrong for Facebook's Libra, *Financial Times*: www.ft.com/content/6e29a1f0-ef1e-11e9-ad1e-4367d8281195

TREND 7
CLOUD AND EDGE COMPUTING

The One-Sentence Definition

Cloud computing means storing and processing data on other people's computers (e.g. data centers) via a network (e.g. the internet), which gives companies the ability to store massive amounts of data and process it in nearly real time. Edge computing refers to the processing of data on devices such as smart phones (which are getting more powerful and therefore no longer need to outsource processing to the cloud).

What Are Cloud and Edge Computing?

To put it in the simplest terms, the cloud is "other people's computers." With the rise of cloud service providers such as Amazon, Google, and Microsoft, it's no longer necessary or even desirable to host all of your vital IT infrastructure within your organization's digital walls.

Moving operations to the cloud means you cut down on the overheads needed to maintain and operate all of your systems, software, and data. Cloud providers host all of the tools for you, allowing you to access them wherever and whenever you need to. Not only does this mean you take advantage of their expertise at maintaining and updating tools, you also benefit from their world-class security and support facilities. You also get access to the huge amount of computing power and storage resources that cloud service providers have at their disposal – the amount of computing resource at your disposal can be scaled up or down, as demand for your own service changes.

Of course, the providers themselves benefit too – they no longer have to struggle with offering support to a customer base that could be using millions of different permutations of software and hardware. They also get access to data about how, where, and when we are using their platforms, allowing them to further tailor their service offerings to fit their customers' needs (see also Trend 18). And, of course, they get to charge us directly, using a subscription model.

Edge sits at the other end of the scale – rather than far away in remote data centers, edge computing happens up-close-and-personal on the frontline of your business operations. Rather than send every piece of information collected by cameras, scanners, handheld terminals, or sensors to the cloud to be processed, edge devices carry out some or all of the processing themselves, at the source – where the data is collected.

Imagine an artificial intelligence (AI)-equipped security camera, for example, equipped with computer vision (see Trend 12) capabilities, keeping watch after hours at an office building. Of all the data it collects, 99% is likely to be worthless imagery of an empty room or corridor. If all of that data has to be sent to the cloud for processing before it can be acted on, not only is bandwidth wasted, but there will also be a delay if an alarm needed to be raised when the data indicates that something out of the ordinary has been detected.

How Are Cloud and Edge Computing Used in Practice?

Millions of us already use cloud software and services in our everyday lives. We use them when we connect to web-based email systems, store photos and videos on online albums, and back up our files to Google Drive or Dropbox.

We're also used to using them when we open as-a-service software packages such as Office 365, Google Documents, or Adobe Creative Suite. Individual components of those packages are kept up to date by the teams administering them, and we can upload our work to the cloud so it can be accessed from any computer or device that we sign into.

Most of our activity on social media is carried out in the cloud, too. We upload images, videos, and text to the servers maintained by the service providers, and we also take advantage of their computing power if we apply filters to our images and edits to our videos.

Many of the apps we use on our phones to carry out day-to-day activities such as ordering cabs, checking train times, or booking movie tickets carry out their activities in the cloud.

Sometimes this will be done in the "private cloud" – where servers are physically owned and operated by the service providers themselves. Sometimes it is in what is often called the "public cloud" – where server space and computing resources are leased from third parties which specialize in providing cloud services, such as Amazon Web Services, Microsoft Azure, or Google Cloud.

As far as public cloud providers go, Amazon has consistently been the market leader in providing cloud services for business for the last three years.[1] Its AWS service includes tools for data storage and processing, analytics, deployment, software development, project management, and Internet of Things functionality.

Many of the most popular consumer-facing cloud apps, such as Netflix[2] or Spotify,[3] rely on public cloud infrastructure to serve their customers, as establishing their own server, processing, and storage facilities within reach of their widely dispersed customer bases would be prohibitively expensive.

The same principles apply to businesses that want to move their operations to the cloud. Platforms exist which allow everything from customer services, inventory management, recruiting and HR, design, retail, and shipping to be handled by cloud providers, "as-a-service."

- **Salesforce Marketing Cloud** is software which lets businesses move all their online, email, and social marketing operations into the cloud, where they can take advantage of advanced analytics and AI-driven recommendation engines to collect and process customer information, and more precisely target marketing campaigns.

- **Evernote** is another business that has been built on the principle of giving customers access to a cloud-based take on a very simple concept – the sticky note. Users can take notes and snippets of information, such as pictures, videos, or voice, and store them in the cloud where they can be accessed from any of their devices or quickly shared with colleagues or friends.

- **American Airlines** worked with IBM to develop a cloud solution which would give customers more flexibility when they needed to rebook flights due to cancellations or service disruption.[4] While customers affected by these circumstances would generally have a seat automatically assigned to them on the next available flight, those with more complex needs would have to contact the airline directly to discuss options. The company developed a cloud-based app which gave passengers all the data they needed to make a choice, based on all of the options available to them.

- Online retailer **ASOS** uses Microsoft cloud services to offer its customers personalized shopping experiences and recommendations. It stores user profiles and customer data in the cloud where it can be quickly accessed and used to determine how relevant particular products are to customers as they are browsing the site.[5]

- Insurers **Aviva** use a cloud system to store and analyze telemetry data taken from drivers' mobile phones, which it uses so it can base its quotes on the behavior of individual drivers.[6] This means it can quote more efficiently and provide more affordable premiums for safer drivers. The system requires data storage and processing capabilities that would have been too costly to implement on-site, in the scalable manner that it required.

Cloud also means companies can offer their employees "virtual desktop" environments which they can access from any device, wherever they are. Rather than allowing staff to simply download business software and data straight to their own machines (with all the security risks that could pose) the apps and data are stored on private cloud services and accessed through a virtual desktop.

Edge computing is about leveraging the processing power of devices close to the source of data that is being collected, to save on bandwidth used by sending it to the cloud, and processing that needs to be done once it is there.

- A good, basic example is online games that are played on a local console. These require only a small subset of the data generated by the game to be sent to the cloud – generally only the data which will affect other players in the game. Meanwhile, the majority of the data processing is done on the user's own console, and the video data that this generates is visible only to them, on their own display.

- A more complex use case can be found in the automated vehicles that are quickly becoming a reality. These vehicles rely on sensors to detect any danger of collision and take evasive action before it happens. In life-or-death scenarios like this, when vehicles are traveling at high speed, it wouldn't be sensible to expect to send data to a location where it would be processed in the cloud before making a decision on whether or not a danger was present, and relaying it to the computers which control the vehicle's motors. In these cases, data gathered by cameras and radar/lidar (light detection and ranging) sensors will be analyzed before it leaves the vehicle. Only relevant data will be sent back to the cloud, where it will be used to make less time-critical decisions, such as those affecting route planning, fuel optimization, and vehicle performance.

- The smart cities concept (Trend 5) is about using technology to improve services and utilities in urban environments, and is a fertile ground for the deployment of edge computing initiatives. Systems for monitoring and reacting to traffic movement and congestion can rely on image processing built into cameras and react to changing situations by altering signals or activating temporary speed restrictions. Systems incorporating CO_2 monitors could reroute vehicles when emission levels become unhealthy in a particular area, and waste disposal facilities can send an alert when sensors detect they are at capacity, so they can be emptied more efficiently. Without the ability for a lot of this work to be done at the edge, central servers could become overloaded and bogged down with the number of different systems that are constantly sending and requesting data.

- In industry, edge is quickly becoming popular in environments where there is little or no access to online services. In remote mining and offshore oil facilities, data analytics can be handled locally in order to make minute-to-minute decisions based on data generated locally.

- Manufacturing plants also use edge analytics to understand how their equipment is operating, and to enable them to carry out predictive maintenance – understanding when mechanical problems are likely to occur, and fixing them before they do.

Key Challenges

Perhaps the first consideration that has to be taken into account is the cost. As one of the primary drivers towards cloud solutions is the reduction in cost by reducing the need for on-premises infrastructure, it's vital to consider the additional costs that cloud platforms require in terms of support and scalability. If you're offering services to customers via the cloud, the usage will spike as your userbase grows, leading to an increase in cost.

Security is always a major challenge with IT operations, and while cloud solves many of them – for example, it prevents thieves from physically stealing data by taking devices from your premises – it causes some others.

An important one to consider is that the principle that cloud services can be accessed from anywhere means they can theoretically be stolen from anywhere, too. Systems that are used to authenticate access to the services – such as login details, passwords, and tokens – need to be carefully audited to ensure they are managed in a secure way.

This leads to a third challenge, in that you will be becoming reliant on a third party and putting some important factors, such as data security, in their hands, when moving services to the cloud. For this reason it's essential to thoroughly research whatever cloud service providers you are considering using, and understanding how they manage issues such as privacy and compliance. In certain jurisdictions, such as Europe where data use is governed by the General Data Protection Regulation, it's likely you can still be held legally responsible if data that your organization is responsible for is mishandled by third-party service providers.

Migrating to cloud also means you become reliant on the provider for ensuring your own continuity of service. Providers frequently change the products and services they offer, as well as the level of support they provide, and if a service you rely on is withdrawn or its functionality is altered, you may be left scratching your head about how to continue filling the needs of your own customers!

With edge computing, the challenge is often ensuring that in the mission to save bandwidth, you are not ignoring or discarding important data that could have other uses, aside from what is happening in the edge device.

Taking the example of an automated vehicle, it may seem that the transfer of millions of images gathered while a car travels along an empty road would be pointless. However, data about the condition of the road and the environment the vehicle is passing through could still be useful for regulating how other vehicles behave when making the same journey. Balancing the value of data with the bandwidth and storage that will be consumed by sending it to the cloud is vital.

How to Prepare for This Trend

It's important to determine the cost in terms of infrastructure, IT support, and data bandwidth consumed by IT operations across the organization, department by department, and use by use. This means you can make informed decisions on whether cloud migration makes commercial sense for any particular business process.

As with any deployment of new technology, it makes sense to start by looking for "quick win" use cases. These are opportunities to move smaller-scale processes or operations to the cloud, in what will be simple deployments, and when the usefulness of the move can be quickly assessed. These can also help make a business case for transitioning larger processes to a cloud deployment.

Following that, it's useful to familiarize yourself with the services offered by the main cloud computing providers – the big three are generally considered to be Amazon Web Services, Google Cloud, and Microsoft Azure. This helps to gain an understanding of whether their services and tools are in-line with your business needs.

Beyond that, switching to working in a cloud environment can require some fairly radical changes in mindset on the part of those whose job it is to manage and deploy your infrastructure. When they are no longer so concerned about building and maintaining in-house IT systems, they will be free to take a more strategic, higher-level overview of how technology is being used to meet goals and drive performance. This means becoming familiar with the tools that are on offer and the possibilities they unlock.

You will also have to ensure that you have the relevant expertise in security, data management, and automation in-house in order to be able to effectively assess, access, and operate the services offered by cloud providers.

It's also worth considering how the existing external services you use for critical IT operations perform – for example, your internet service provider. When the majority of your technical work is being done in-house on your own local area network, you may be resilient to drops in service quality. However, when you're relying on a 24/7 connection to the cloud in order to serve your customers, any drop in quality could worsen the experience of your customers. Is this something you can afford to risk and, crucially, what contingency plans can you put in place if it happens?

Notes

1. Top cloud providers 2019: AWS, Microsoft Azure, Google Cloud; IBM makes hybrid move; Salesforce dominates SaaS: www.zdnet.com/article/top-cloud-providers-2019-aws-microsoft-azure-google-cloud-ibm-makes-hybrid-move-salesforce-dominates-saas/
2. Netflix on AWS: https://aws.amazon.com/solutions/case-studies/netflix/

3. Switching clouds: What Spotify learned when it swapped AWS for Google's cloud: www.techrepublic.com/article/switching-clouds-what-spotify-learned-when-it-swapped-aws-for-googles-cloud/
4. American Airlines: www.ibm.com/case-studies/american-airlines
5. Online retailer uses cloud database to deliver world-class shopping experiences: https://customers.microsoft.com/en-gb/story/asos-retail-and-consumer-goods-azure
6. UK Insurance Firm Uses Mobile App and Cloud Platform to Track Driving Behavior https://azure.microsoft.com/en-gb/case-studies/customer-stories-aviva/

TREND 8
DIGITALLY EXTENDED REALITIES

The One-Sentence Definition

Extended reality (XR for short) – which encompasses virtual reality, augmented reality, and mixed reality – refers to the use of technology to create more immersive digital experiences.

What Are Digitally Extended Realities?

Let's start by breaking down the different types of XR:

- **Virtual reality** (VR) means using computer technology to fully immerse the user in a simulated digital environment, to the extent that they feel like they're physically in that environment. VR typically works via special headsets or glasses, like the Oculus Rift, HTC Vive, or Samsung Gear VR headsets. VR technology has advanced enormously in recent years and can now offer incredibly realistic digital experiences, such as walking on the moon, or wandering around eighteenth-century Venice. (Just think how much more interesting history lessons might be for the children of the future!) Entire theme parks are now being designed around VR experiences – China's VR Star Theme Park is one prominent example.

- **Augmented reality** (AR) is rooted in the real world, not a simulated digital environment. With AR, information or images can be overlaid onto what the user is seeing in the real world – the popular Pokémon GO game, where users could "see" Pokémon characters on the street via their smart phone, is a well-known example. AR can be delivered via smart phones, smart glasses, tablets, web interfaces, smart screens, or smart mirrors. Because AR is rooted in the real world, the experience is less immersive than VR but offers amazing opportunities to augment the world we see around us.

- **Mixed reality** (MR) is an extension of AR that brings the virtual and real worlds together, creating a more connected experience in which the virtual and real elements can interact. With AR, users can't interact with the information or objects being overlaid onto the real environment; with MR, they can. The user can play around with virtual elements like they would in the real world, and the 3D digital content responds and reacts accordingly. For example, using gestures, you could physically turn an object around to inspect it from all angles. For MR to work, the user has to have an MR device, like Microsoft's HoloLens, which, among other things, turns apps into holograms that the user can touch and move around. It's worth checking out a video of the HoloLens[1] or Magic Leap[2] in action to best understand how MR works.

All three XR technologies offer exciting and entirely new ways for people to experience the world around them – and for businesses, new ways to connect and engage with customers and improve business processes.

If it sounds like I've got lost in the plot of a sci-fi movie, think again. As you'll see in this chapter, XR technology is already finding very real applications in our world, and is likely to dramatically change the way we interact with technology. In fact, mobile-based AR experiences (such as the Pokémon Go app) generated over $3 billion in global revenue in 2018.[3] No wonder, then, that Accenture's *Technology Vision 2018* survey found that more than 80% of executives believe XR will create a new way for businesses to interact and communicate.[4]

How Are Digitally Extended Realities Used in Practice?

The worlds of gaming and entertainment were obvious early adopters, but XR technology is now being used in a range of industries and organizations – from training surgeons and soldiers, to selling the latest luxury vehicles. As you'll see in this section, many brands are already tapping into XR to create memorable, immersive experiences for their customers and employees.

Of the three XR technologies, MR is the newest and therefore the least widely developed, so the following examples naturally focus more on VR and AR.

Boosting Brand Engagement

Thanks to XR, brands are able to design fun, novel experiences that help to create a buzz around their brand.

- **Pepsi** created an incredible AR-enabled display in a London bus shelter, which stunned commuters by overlaying eye-popping images onto the real-life street in front of them. These images included a meteor crashing into the ground, a tiger padding towards them, and a large tentacle popping up from underneath the paving slabs!

- **Mercedes** created a sleek virtual experience of driving the latest SL model down California's beautiful Pacific Coast Highway.

- **Uber** installed an AR experience at Zurich train station that immersed passers-by in adventures such as petting a tiger in the jungle. A video of people interacting with the experience has had more than a million YouTube views.[5]

- **Burger King** has created an AR-enabled app feature that lets Burger King fans "burn" the ads of rival fast food chains – in exchange for a free Whopper. Using the "Burn That Ad" feature in Burger King's app, users point their smart phone at the competitor's ad and watch it go up in impressive flames, after which the ad is replaced with an image of a Whopper, with a link to claim a freebie burger.

Letting Customers Try Before They Buy

With AR, customers can try a product or get to know a brand's product range from the comfort of their own home.

- **Ikea** has created an AR app that lets customers bring Ikea furniture to life in their own home. Called Ikea Place, the app allows you to scan your room and place Ikea objects in the digital image of that room, creating a whole new look.

- Family-owned retailer **Tenth Street Hats** offers an AR solution that lets customers try on hats at home. Users can see how different hats look on their head at any angle, and take pictures of themselves wearing the hat.

- The **Dulux** Visualizer tool lets users scan their room and virtually "paint" it in any color.

- Cosmetic companies like **L'Oréal** and **Sephora** are using AR tech to allow customers to try out different beauty looks before they buy.

- E-commerce platform **WatchBox**, which specializes in preowned luxury watches, has created an AR feature in its app that lets customers "try on" luxury watches before they purchase. The watch in question digitally appears on the customer's wrist in an exact representation of the watch's size, shape, and dimensions.

- **Gap's** Dressing Room app lets shoppers input their body dimensions and try on clothes virtually in a simulated Gap changing room.

- **BMW's** i Visualizer AR tool allows users to see, customize, and interact with a full-size Beemer.

Enhancing Customer Service

Finding ways to make customers' lives easier gets a whole lot more interesting when you add VR and AR into the mix.

- The passenger app for London's **Gatwick Airport** has won awards for its creative use of AR technology.[6] Passengers can use the app to navigate through the busy airport.

- Similarly, ride-sharing app **Didi Chuxing** has an AR feature that guides passengers through busy buildings to find their exact pick-up location.

- For anyone who's ever trudged to the DIY store and realized too late that they left the tape measure at home, **Lowe's** has created the Measured by Lowe's virtual tape measure that uses AR to turn your smart phone into a tape measure.

Making Workplace Learning More Effective

XR offers new ways to immerse employees in a situation, thereby enhancing the learning experience.

- For those who dread public speaking, Oculus's **VirtualSpeech** VR tool provides immersive training that helps people deliver better sales pitches, network more effectively, become a more confident speaker, and more.

- The **US Army** is harnessing AR to improve soldiers' situational awareness, using an eyepiece that helps them precisely locate their position, locate others around them, and identify whether they're a friend or a foe.

- **Children's Hospital Los Angeles** partnered with VR specialists BioflightVR and AiSolve to create VR-based training scenarios for pediatric surgeons.[7] The simulation is so detailed, the creators scanned and created 3D representations of the hospital's real nurses, so that trainees' virtual experience would match the real-life operating theater.

- A team from the **University of Virginia** has created a VR-based classroom where teachers can test their delivery and classroom management, and receive instant feedback.[8]

- Law enforcement officers in **New Jersey** are using a system that allows them to train for various scenarios, ranging from routine traffic stops to being shot at.[9]

Improving Other Organizational Processes

Although most of the industry use cases focus around marketing and training, VR, MR, and AR technology is helping to enhance a wide range of other processes and functions.

- With XR, every characteristic of a component or manufacturing process can be simulated and tested, without having to build expensive working prototypes – potentially a game changer for many manufacturers. In aircraft design, for example, **Boeing** and **Airbus** both extensively use simulated digital spaces to design and test new features and models. Similarly, **Ford** is using Microsoft's HoloLens headset to design cars in mixed reality – indicating the huge potential for MR, once the technology becomes more widespread.[10]

- Facial recognition technology (see Trend 12) is being increasingly used by law enforcement agencies around the world. And when coupled with AR eyewear, officers can cross reference faces in a crowd to identify criminals in real time. **Police officers** in public security bureaus in China are already using AR glasses to check faces against the national database.[11]

- Even recruitment can be enhanced through XR. Foodservice company **Compass Group** is a huge employer with over 500,000 employees, but it's hardly a household name – and this lack of brand awareness made attracting talented graduates more challenging.[12] To overcome this, the company created a VR experience for campus events that lets students take a virtual tour of the workplace and participate in a video interview.

Key Challenges

Accessibility and availability is one obvious obstacle to overcome, given that XR headsets can be pricey, bulky, and clunky to use, which puts off both individual consumers and corporate users alike. But the technology will become more common, affordable, and comfortable to use – which will only increase the chance of widespread use in businesses. AR is ahead of VR in this respect, and many slick AR experiences can be designed for use on a basic smart phone or tablet.

So, setting aside the technology issues, perhaps the biggest challenges for those wanting to adopt XR technology come from less obvious factors, such as privacy and the potential mental and physical impact of highly immersive technologies.

Let's start with privacy. With XR technology, intimate behaviors – what we look at, where we go, what we do, even to some extent what we think and feel – can be tracked in great detail. What happens to this highly personal information, and how can we be sure it's not used in unethical ways? Personal information like this is open to misuse, theft, and manipulation, and we could see identity theft taken to extreme new levels. Forget about criminals stealing your credit card info; imagine them creating a digital doppelganger of you, and making that doppelganger do something embarrassing or illegal in the digital world.

Then there's the potential impact on users' mental health. The full implications for those who spend time in XR isn't yet understood. Overdependence is a major concern, and the more time people spend in XR, the harder it may become to separate the real from the non-real. Social media has already created discrepancies between some people's real lives and the #blessed version of themselves they present online – will XR widen this gap? Very likely. Imagine someone who spends a lot of time in a "perfect" online world, and how they would react when reminded of the messiness of the real world (war, poverty, and pollution, to name just a few issues that continue to make the real world less than perfect)? Is that person more likely to retreat further into their virtual paradise, or engage with social issues and work to make the real world a better place? Most of us would bet on the former.

This has some concerned that heavy XR users may become increasingly disengaged from real life, and that new mental health disorders may emerge. (If this sounds far-fetched, consider that, in 2019, the World Health Organization recognized gaming addiction as a mental health disorder.[13] Consider the risks as virtual environments become more and more immersive.) Another concern is that online bullying may become more extreme in the virtual world. After all, instead of flinging anonymous insults at people, trolls could in theory physically bully and intimidate their victims in the digital space.

As well as mental health concerns, there are implications for our physical health and safety to consider. For example, an AR headset that overlays information on the real world could create dangerous distractions for pedestrians and drivers – especially if the technology is vulnerable to attack. In the future, if many of us are walking around with AR glasses, hackers could overlay mischievous or downright terrifying images onto the real-life street to cause panic and unrest in communities.

What's more, there are physical side effects from using XR headsets for too long, and most manufacturers recommend users take regular breaks to avoid side effects like loss of spatial awareness, dizziness, disorientation, nausea, eye soreness, and even seizures.[14]

All these concerns are supported by a recent report from Accenture that emphasizes the physical, mental, and social risks associated with XR tools – risks it says are far greater than those resulting from existing technologies.[15] And while some of these concerns may seem beyond the paygrade of a manager considering, say, an AR-based marketing campaign, it goes to show that businesses must approach XR technology in a responsible way – with privacy, security, and ethics squarely in focus. As with any innovative technology, it pays to be vigilant about the potential hazards so that you can harness its true potential and reap maximum benefits.

How to Prepare for This Trend

The impact of XR will vary from organization to organization, but, as a general rule, it's fair to expect that the use of XR will become more and more common across all industries – particularly when it comes to marketing, customer engagement, and workplace learning. It makes sense, then, for companies to start thinking strategically about how they might capture the benefits of XR.

The following questions will help you kickstart this process of consideration and begin to define your priorities:

- **How is XR already being used in your industry?** A good starting point is to look at what's already going on in your industry. For example, is VR making waves in your particular sector? Or are some of your competitors creating AR apps for consumers?

- **From a marketing and branding perspective, how might XR help differentiate your products, service offering, or brand from the competition?** Think about what makes your company different from others in your industry, and how VR, AR, and MR could help reinforce your unique proposition. (Think how Burger King's flame-grilled flavor informed the company's "Burn That Ad" campaign.)

- **How might XR help to improve your internal business processes, such as training, recruitment, or manufacturing?** Reducing costs, improving quality, and creating operational efficiencies are just some of the benefits of using XR in-house.

- **Do you have the necessary technical skills in-house or would you need to partner with a provider who specializes in XR?** Although dedicated in-house teams will become more common, for now, most companies have no choice but to partner with external specialists. That said, if the use of XR is absolutely critical to achieving your strategic goals, then it's wise to focus on building in-house capabilities.

- **Do you have the digital content you need?** Effective VR, AR, and MR requires content. Depending on what you have in mind, this may include training materials, product images and specifications, operational procedures and instructions, etc. If you don't already have the content you need, how will you go about creating it?

- **What hardware will users need?** Many AR and some VR experiences are designed for the ubiquitous smart phone, and all the user needs to do is download an app to get started. Meanwhile, a fully immersive VR experience may require dedicated equipment, like a headset. When considering hardware, think about the kind of experience you want to create and who you're aiming at.

Notes

1. HoloLens 2 AR Headset: On Stage Live Demonstration: www.youtube.com/watch?v=uIHPPtPBgHk
2. Mixed Reality demo showing a whale jumping: www.youtube.com/watch?v=LM0T6hLH15k
3. For AR/VR 2.0 to live, AR/VR 1.0 must die: www.digi-capital.com/news/2019/01/for-ar-vr-2-0-to-live-ar-vr-1-0-must-die/
4. Technology Trends 2018: www.accenture.com/dk-en/insight-technology-trends-2018
5. Augmented reality experience at Zurich main station: www.youtube.com/watch?v=bCcvEVyAXQ0
6. Gatwick's Augmented Reality Passenger App Wins Awards: www.vrfocus.com/2018/05/gatwick-airportsaugmented-reality-passenger-app-wins-awards/
7. How VR training prepares surgeons to save infants' lives: https://venturebeat.com/2017/07/22/how-vr-training-prepares-surgeons-to-save-infants-lives/
8. Using VR To Help Support Teacher Training; Huffpost: https://www.huffpost.com/entry/using-vr-to-help-support_b_10114136?guccounter=1
9. Virtual reality helps reinvent law enforcement training, CBS News: https://www.cbsnews.com/news/virtual-reality-law-enforcement-training/
10. Ford is now designing cars in mixed reality using Microsoft HoloLens: https://techcrunch.com/2017/09/22/ford-is-now-designing-cars-in-mixed-reality-using-microsoft-hololens/
11. The Amazing Ways Facial Recognition AIs Are Used in China, Bernard Marr: www.linkedin.com/pulse/amazing-ways-facial-recognition-ais-used-china-bernard-marr

12. How AR and VR are changing the recruitment process: www.hrtechnologist.com/articles/recruitment-onboarding/how-ar-and-vr-are-changing-the-recruitment-process/

13. Video game addiction now recognized as a mental health disorder by the World Health Organization, *Daily Mail*: www.dailymail.co.uk/sciencetech/article-7079529/Video-game-addiction-recognized-mental-health-disorder-World-Health-Organization.html

14. Here's what happens to your body when you've been in virtual reality for too long: www.businessinsider.com/virtual-reality-vr-side-effects-2018-3?r=US&IR=T

15. A responsible future for immersive technologies: www.accenture.com/us-en/insights/technology/responsible-immersive-technologies

TREND 9
DIGITAL TWINS

The One-Sentence Definition

A digital twin is a digital copy of an actual physical product, process, or ecosystem that can be used to run virtual simulations, using data to update and change the digital copy to reflect any changes in the real world.

What Is a Digital Twin?

The term "digital twin" was first used by Michael Grieves at the University of Michigan in 2002,[1] but the concept itself goes back further. NASA pioneered the model of working with digital models of real-world systems during its Apollo missions, and having accurate simulations, based on real-world data, is credited with helping it to recover its astronauts safely back to Earth following equipment failure on Apollo 13.

The emergence of the Internet of Things (IoT – see Trend 2) and artificial intelligence (AI – Trend 1) has meant that this has become affordable for a far greater range of businesses and organizations. As everything from everyday items like watches and fridges to industrial machinery operating in manufacturing facilities can now be collecting and sharing data, anyone can use this data to build digital models.

The idea is that it lets us see what might happen if we make adjustments that might be too expensive, dangerous, or uncertain to try out in a real-world scenario. By altering the variables under which the digital twin is operating, the changes can be observed in the digital world without putting money or safety at risk.

For a simple example, think of a shop. A shopkeeper could alter staff levels, prices, stock levels, contractual expenses, variable expenses such as lighting and refrigeration, and customer facilities within his or her digital model, and monitor the effect this will have on metrics such as turnover, profit, and customer loyalty.

To do this, the model would need to be fed with accurate data on how the shop operates in the real world, so it can get to "understand" its relationship to the metrics. Large retailers are experimenting with automating the capture of this data, using computer vision (Trend 12) to monitor stock levels and expiry dates. The more advanced implementations of this technology will automatically learn how to keep more accurate track of the variables it is monitoring, meaning that the technology can also be thought of as an implementation of AI and machine learning.

As well as driving efficiency and growth within existing operations, they also enable new products or services to be designed and prototyped entirely from scratch. Building within a digital twin environment, informed by real-world data, means designers and engineers will have a more deeply informed understanding of how their end product will work when it's unleashed on the world.

Businesses can also use the tools to monitor their impact on the environment, by building a more accurate picture of the energy they use and the emissions they create, and how these results can be affected by altering variables such as how heating and lighting are distributed around facilities.

As Thomas Kaiser, SAP's senior vice president of IoT, put it, "Digital twins are becoming a business imperative, covering the entire life cycle of an asset or process and forming the foundation for connected products and services. Companies that fail to respond will be left behind."

Digital twins rely on processing taking place both in the cloud and at the edge (Trend 7). The data that informs the model is collected at the edge by scanners, sensors, or humans working on terminals, while the model simulation runs in the cloud, meaning it can be accessed and used from anywhere.

In 2020, 62% of respondents to a Gartner survey said they are in the process of establishing digital twin technology, or plan to do so in the next year.[2] Recent research by analysts MarketsAndMarkets has found that the value of the market for digital twin solutions is set to grow from $3.8 billion in 2019 to $35.8 billion by 2025, with the largest adopters being the healthcare, automotive, aerospace, and defense sectors.[3]

They are set to become a major element of the IT infrastructure of an increasing number of companies in the near future, meaning that data-driven decision-making will become increasingly ingrained in business, leading to widespread growth. Overlooking the impact they could have on your organization at this point is likely to be a bad business decision.

How Are Digital Twins Used in Practice?

Since proving their effectiveness during the Apollo program, **NASA** has continued to refine their digital twin technology. Today, it is used to build an in-depth picture of vehicle operations, maintenance history, and safety records.

It defines its model of digital twin as an "integrated, multiphysics, multiscale, probabilistic simulation of an as-built vehicle or system that uses the best available physical models, sensor updates, fleet history, etc., to mirror the life of its flying twin.

"The digital twin continuously forecasts the health of the vehicle or system, the remaining useful life and the probability of mission success."[4]

Digital Twins in Motor Racing

Another pioneering use of the concept – from before the term became widely used and it was just considered an advanced form of simulation – can be found in **Formula One** racing. Detailed models are built using data from sensors attached to cars, and on tools used during pit stops, which mean the effects of minute changes can be measured and assessed before a race.

Dr Peter van Manen, a former managing director of McClaren Applied Technologies, said "Formula One is all about time management. Every second counts so when you can shave them off by learning key insights about the inner workings of your car, it really matters. Digital twins are not going to be perfect straight away – they're a bit like a puppy at Christmas – it's great but you have to keep taking care of it if you want to reap the benefits."[5]

Those early examples often concern virtual twins built to simulate particular real-world objects – spacecraft or cars. Today, the concept has expanded to cover simulations of processes, entire organizations, or even ecosystems.

GE offers a service called Digital Wind Farm that allows wind farm operators to understand the optimum configuration of variables for individual wind turbines, before a penny is spent on construction. By tailoring the output to take into consideration factors such as the location and environment a turbine will be operating in, performance tweaks in the digital world cut down on the need for expensive assessment and alteration work after a turbine is up and running.

It says that one major customer was able to make savings of $2,500 per megawatt of energy generated, by cutting the need for expensive interventions. It used existing data on wind energy generation combined with real-time insights on mechanical and electrical components.[6]

Ganesh Bell, GE's chief data digital officer and general manager of software and analytics, says, "For every physical asset in the world, we have a virtual copy running in the cloud that gets richer with every second of operational data."

In **healthcare**, capability is being developed that can be considered to be building digital twins of people. By monitoring data gathered by smart devices such as watches, as well as specialist sensors, hospitals may soon be able to leverage the same principles applied in industrial automation. The concept is being piloted at GE's Health Innovation Village in Finland.[7]

Other implementations of **human digital twins** include those created for China Central Television's Spring Festival Gala coverage in 2019. Human hosts were joined on-screen by AI copies of themselves, using AI to simulate their speech, personalities, and body language. In the future, creating digital twins of real people could enable us to create virtual people to fill just about any role we can think of. This could include scary possibilities like "resurrecting" dead people.

And the digital twin trend is clearly highly relevant to the **smart city** concept, which is quickly growing in popularity. One of the most extreme examples is the complete digital copy that has been developed of the city of Singapore by the nation's National Research Foundation.

It gathers information about demographics, climate, land use, traffic, public transport, and building infrastructure that is used by city planners and those responsible for providing civic amenities to prototype and deploy their work.

The tool, simply known as **Virtual Singapore**, is used for improving accessibility, simulating emergencies in shopping areas or stadiums, deciding on the placement of facilities, such as foot bridges, and monitoring the value of green or sustainability initiatives, such as solar panels and cycle routes.

On an even larger scale, digital twin concepts are heavily used to model whole geographic ecosystems for the purposes of predicting and reacting to natural disasters. Microsoft says that it provides tools that are used by governmental organizations and non-governmental organizations to process satellite imagery, weather stations, calls to emergency services, and social media data to react to incoming threats such as hurricanes, tsunamis, or forest fires.[8]

SkyAlert is a system used in Mexico that monitors custom-made sensors to give citizens up to two minutes' advance warning that an earthquake is likely to strike – vital moments that could make the difference between life and death.[9]

Key Challenges

The usefulness of a digital twin depends on three factors – the quality of the data it is built on, the risks it poses to security, and the task it is being used for.

Data Quality

Firstly, bad data will inevitably lead to digital twins that provide bad predictions. In some ways, the solution to this challenge is built into the concept itself, as the most useful digital twins will gather data directly from the real world themselves, using scanners, cameras, and measuring devices.

But even so, care must be taken to avoid bias in the data – if you're only collecting data across a sample of your output, is that data representative? And are the scanners and other devices which live at the edge accurate enough to provide repeatable, actionable results?

A digital twin built to assess the operation of a customer service channel – such as a customer support chatbot – would be trained on real, anonymized customer behavioral data, gathered from their previous interaction. With a global operation, customers in different countries and regions may have different expectations, meaning similar interactions with the chatbot would lead to different outcomes. Care has to be taken that the insights discovered through use of the digital twin are relevant in specific, individual situations.

Security

Secondly, being by its nature divorced to some extent from the "real world," the risks posed by digital twins seem like they may be less worrisome than is the case with other tech trends.

However, security still has to be taken seriously. Although customer data will generally be anonymized, once its absorbed into the twin's system in the cloud, you're not exposing anything personal, but it still may be commercially sensitive.

Any new data connection between a business in the real world and a virtualized twin in the digital world should be considered a potential point of weakness, and care has to be taken that it doesn't make a system vulnerable to human error, or an opening for an attacker.

The Task in Hand

Selecting the right use cases for digital twins, and ensuring that they work towards your overall digital and analytics strategies, is the third challenge. As with any new breakthrough technology, there can be an impulse to dive in headfirst and start trying to apply it to everything. Often this leads to the most obvious use cases turning out not to be the best strategic first choices. At the best this leads to wasted time and effort. At the worst it can lead to loss of faith in the technology from key stakeholders, including those with the power to pull the plug on future projects involving it.

How to Prepare for This Trend

Successful digital twin applications can be accomplished by working through solutions to all of the challenges mentioned above.

Starting with data, you will have to ensure that your pipeline for collecting, validating, and storing data is up to date and efficient. Most likely, as the twin becomes a tool for ongoing use, you will need to have a process for continually evaluating and adapting your data pipeline.

As the twin grows and is applied to an increasing number of business cases, the volume as well as the speed of the data it must ingest will increase. Building scalable solutions to these problems will be fundamental to the organization's overall business analytics strategy, as well as the specific implementation of the digital twin.

Likewise, a digital security strategy should be in place covering all data gathering, analytics, and storage activity. To ensure that it is tailored towards security, you'll have to be careful that the edge nodes of the network – where data is collected at source – are not vulnerable to becoming inaccurate due to human error (for example, during manual calibration) and also that data streams don't become targets for hackers.

When it comes to ensuring that digital twins are being deployed to usefully tackle problems and in support of an overall analytics strategy, make sure that the technology is being applied to problems which are appropriate – meaning they can be solved by better understanding the data you have available – and in line with overall business objectives.

This means you should be able to demonstrate a clear link between the predictions you're hoping to generate and the metrics and performance indicators you want to improve.

Digital twins might not always be cheap to develop – divorced (though not isolated) from the real world as they are, people with the skills needed to design and deploy them may not be readily available in an organization. This means they must serve a clear business need and ultimately deliver return on the investment put into building them.

As with other forms of advanced analytics, identifying "quick win" applications is generally a good start. These are smaller-scale, simplified deployments where the usefulness of the technology can quickly be either demonstrated or disproved. If it works out, as well as starting to get an understanding of the work involved, it becomes easier to argue for its use in larger-scale, more expensive applications.

Notes

1. Identical Twins: www.asme.org/topics-resources/content/identical-twins
2. Gartner Survey Reveals Digital Twins are Entering Mainstream Use: www.gartner.com/en/newsroom/press-releases/2019-02-20-gartner-survey-reveals-digital-twins-are-entering-mai

3. Digital Twin Market by Technology, Type (Product, Process, and System), Industry (Aerospace & Defense, Automotive & Transportation, Home & Commercial, Healthcare, Energy & Utilities, Oil & Gas), and Geography – Global Forecast to 2025: www.marketsandmarkets.com/Market-Reports/digital-twin-market-225269522.html
4. The Digital Twin Paradigm for Future NASA and U.S. Air Force Vehicles: https://ntrs.nasa.gov/archive/nasa/casi.ntrs.nasa.gov/20120008178.pdf
5. Singapore experiments with its digital twin to improve city life: www.smartcitylab.com/blog/digital-transformation/singapore-experiments-with-its-digital-twin-to-improve-city-life/
6. Renewable Wind Farms: https://www.ge.com/renewableenergy/digital-solutions/digital-wind-farm
7. Healthcare Innovation Could Lead to Your Digital Twin: www.digitalnewsasia.com/digital-economy/healthcare-innovation-could-lead-your-digital-twin
8. Using AI and IoT for Disaster Management: https://azure.microsoft.com/en-gb/blog/using-ai-and-iot-for-disaster-management/
9. Sky Alert: https://customers.microsoft.com/en-us/story/sky-alert

TREND 10
NATURAL LANGUAGE PROCESSING

The One-Sentence Definition

Natural language processing (or NLP for short) refers to the technology that allows computers to understand human language.

What Is Natural Language Processing?

NLP is used to help computers read, edit, and write text – but it also powers computer "speech," as seen in voice interfaces and chatbots (see Trend 11).

If you think about it, so much of the world's information is in the form of natural human language: think of emails, social media posts, text messages, books, spoken conversations, and the like. Traditionally, computers haven't been great at extracting meaning from language, because language is unstructured data (as opposed to structured data of the sort found in data tables and spreadsheets). But, now, thanks to advancement in artificial intelligence (AI) disciplines like machine learning (see Trend 1), computers are able to process and extract meaning from language to an impressive extent. NLP is a subset of AI, but it also relies on big data (Trend 4), since it takes an awful lot of language data to train NLP models and for them to continually get better over time.

Chances are you've already interacted with NLP in one way or another – it's what allows your Alexa or Siri or Google Assistant to understand your requests. Following on from that, natural language generation, or NLG, is what allows Alexa and friends to respond with human-like speech. NLG (also a subset of AI) takes data and transforms it into language that sounds natural, as if a human was writing or speaking. Essentially then, NLP figures out what message is being communicated, while NLG communicates a message. Put both together and we're living in an era where machines can communicate more naturally with humans.

As we'll see in this chapter, the practical applications of NLP and NLG go way beyond smart virtual assistants. The ability to interact with machines through language could transform many everyday processes in homes and organizations across the world. It can offer customers a frictionless experience, for example. It is powering email filters, search engines, translation apps, speech recognition systems, and more. And as the technology improves, we'll experience yet more applications where machines can consume and even match human-generated content.

How does it work? In very simple terms, NLP and NLG involves applying algorithms to extract rules from unstructured language data and convert the data into a format that machines can understand. This process uses analytical techniques such as syntactic analysis (assessing language according to grammatical rules) and semantic analysis (getting at the meaning conveyed by the language). As you can probably imagine, the latter is much more complex than the former.

So, does this mean machines can really *understand* human language? To help answer this question, a linguist at New York University came up with a set of reading comprehension tasks to test computers' ability to really understand text. The test was called GLUE (General Language Understanding Evaluation) and comprised exercises that most humans would find fairly straightforward, such as determining whether a sentence is true based on what was said in the previous sentence. According to the resulting paper, published in 2018, most systems tested didn't do that well – comparable to a reading grade D + for a human student.[1] Then Google introduced a new tool called BERT, which scored much higher – the equivalent of a B–. Several BERT-based neural networks (see Trend 1) began acing the GLUE tests, some of them even outranking human performance.

The BERT-based systems could read as well as humans, if not better. But does that mean they really understand human language, or were they just getting better at taking the tests? Many researchers believe we're still a long way from machines fully understanding all the nuances of language. The truth is, natural language is messy and haphazard and it doesn't always follow perfect rules. The English language, for example, is amass with contradictions, idioms, and words that sound the same but have entirely different meanings. All of this makes life difficult for computers. NLP and NLG systems can model specific tasks extremely well but can't yet cope with the infinite range of linguistic problems that humans solve without thinking. In other words, it's the difference between applied AI (which machines are now very good at) and general intelligence (where machines simply can't compete with the overall intelligence of the human brain) – read more about this back in Trend 1.

Saying that, the technology is advancing at an incredible rate, and many tools exist that can accurately interpret speech and text, derive meaning from it, and even detect the underlying sentiment from what was said. As the technology continues to advance, we can expect machines to get even better at interpreting human language, meaning human–machine interactions will become more fluid.

How Is Natural Language Processing Used in Practice?

Perhaps the best-known examples of NLP and NLG in action come from the digital assistants, like Siri or Alexa, that we've welcomed into our lives. Turn to Trend 11 to read many more examples of voice interface systems; in this chapter, I'll focus mostly on non-spoken communication, such as signing and written text.

Overcoming Communication Barriers

Google Translate is a particularly well-known example of NLP, but let's look at some other ways in which NLP is helping people navigate communication challenges.

- The **Livox** app uses NLP to help people with disabilities communicate. It was developed by Carlos Pereira to help his non-verbal daughter, who has cerebral palsy, communicate with her family.[2] The app is now available in multiple languages.

- The **SignAll** tool can translate sign language into written English, helping individuals who are deaf communicate with people who don't know sign language.[3]

Examples from the Tech World

Many of us are interacting with NLP technology every day without realizing it…

- **Email spam filters** were one of the earliest applications of NLP. But nowadays email providers use NLP for more than just filtering out questionable emails. **Gmail**, for example, can classify incoming emails according to whether they're primary, social, or promotional emails.

- Likewise, **search engines** use NLP to understand your search request and return relevant results. NLP also fuels the predictive search function that anticipates the rest of your search query as you're typing. The same technology underpins **autocorrect and autocomplete** functions in email, word processing, and smart phone applications.

- **Grammarly** is a great example of NLP in action. Since its launch in 2009, the tool has grown to 15 million users,[4] and is one of the top-ranked grammar checkers. You can download the Grammarly Keyboard for mobile devices, download a Grammarly plug-in for Microsoft Office, and add an extension to Chrome – providing a complete spelling and grammar check for Word documents, social media, and email. The AI-based system was trained using examples of correct and incorrect grammar, punctuation, and spelling. And when humans ignore a proposed Grammarly suggestion (perhaps because it's not right in that context), the system learns from that in order to deliver better suggestions in future.

Applying NLP and NLG in Business Settings

You might already be familiar with marketing tools that mine social media for brand mentions and assess underlying sentiment (i.e. whether customers are happy with the brand or not). Such tools wouldn't be possible without NLP. But how else are organizations using NLP and NLG to improve their core business activities?

- NLP can be used to assess the creditworthiness of customers with little or no credit history. For example, the **Lenddo** app uses NLP and text mining to generate a credit score based on thousands of data points from social media and smart phone activity.[5] Tools like this analyze online behavior to unearth insights about the applicant that may predict future activity.

- Swedish bank **Swedbank** deployed Nuance Communications' AI assistant Nina to help customers answer their banking questions. Customers can ask transactional questions via the bank's homepage search feature, using freeform text, and Nina delivers the answers in a conversational tone. Nuance says Nina handles 30,000 queries a month for Swedbank, with a "first-time resolution" rate of 78%.[6] Read other examples of chatbots in Trend 11.

- The **Textio Hire** tool uses NLP to analyze and tweak job descriptions in order to help companies attract the best talent. The tool suggests changes such as cutting biased terms, adding more engaging language, and using phrases that will attract a more diverse candidate pool.[7]

- An NLP-based tool was found to be better at **analyzing risk** in non-disclosure agreements than experienced lawyers. The computers managed an average 94% accuracy rate, while the lawyers averaged 85% accuracy. And the computers were much faster, completing the task in just 26 seconds, versus 92 minutes for the humans.[8]

Transforming Journalism

Aside from reading, machines are getting better at creating content, which means NLP and NLG has the potential to transform journalism forever. Already, many news outlets are using AI-based tools to summarize or generate content automatically and augment the work of human reporters.

- **Bloomberg's** Cyborg tool takes financial reports and turns them into news stories, turning out thousands of articles on companies' earnings reports every quarter.[9]

- **Forbes**, a site that I write for frequently, has a tool called Bertie, which helps reporters generate first drafts for news stories, make headlines more engaging, and find relevant imagery.[10] It's named after the company's founder.

- The **Washington Post** has a robot reporting tool called Heliograf, which churned out 850 articles in its first year.[11] The tool can uncover trends and alert reporters – thereby doing a lot of the behind-the-scenes legwork that goes into news stories.

If writing short-form content like news articles wasn't impressive enough, machines can now write books, too. Turn to Trend 17 (machine co-creativity) to see examples of how machines can now generate entire novels and academic books.

NLP in Healthcare Settings

In the future, could NLP help improve healthcare and deliver better patient outcomes? These examples certainly suggest that's the case.

- NLP has been used to analyze the **risk of heart failure** in patients. In a trial, medical reports of patients that had already been hospitalized were analyzed to predict the likelihood of patient readmission or mortality within the next 30 days. At the end of the evaluation, the NLP model's positive predictive value was 97%.[12]

- **McKinsey** used NLP to enhance clinical benchmarking guidelines by analyzing clinical guidelines from multiple sources, then automatically organizing and classifying the information – resulting in a 60% decrease in the time needed to create clinical guidelines.[13]

- **Nuance Communications** has developed a speech recognition tool called Dragon Medical One, which transcribes clinicians' speech into an electronic health record.[14]

Key Challenges

As mentioned earlier in the chapter, natural language *processing* isn't necessarily the same as natural language *understanding* – where machines exhibit true understanding of all the messy complexities of natural human language in the same way as humans' innate understanding of their language. To reach this point, there are several obstacles to overcome.

For one thing, there are challenges around the language itself – specifically, do we need to develop NLP and NLG tools for every different language, or is it possible to develop a general approach that can be applied to all languages? Certainly there are similarities between languages. But gathering enough data to train such a general model – if it's even possible – would be quite an undertaking.

There are also issues around extracting meaning from a large amount of text or speech (reading an entire book, summarizing it, and pulling out its key themes, for example). To train models on large tasks like this requires a lot of data, computing power, and supervision time – which means, for now at least, NLP and NLG is mostly limited to smaller tasks, such as understanding commands or summarizing information for news articles.

And there are challenges within the analytics process. For example, when text data isn't easily separated into sentence units (if, say, the text is contained in graphics, tables, or notations, rather than neatly flowing paragraphs), it can make it difficult for machines to process the information and extract meaning. But perhaps one of the biggest analytics challenges is training computers to derive *context* from language. Many words have multiple meanings ("bank," "break," and "lie" being just a few examples), and understanding the difference comes down to context. Computers use various methods to tell the difference between meanings, but these models aren't universal. Cracking the problem of context may prove critical in developing true language understanding.

How to Prepare for This Trend

NLP and NLG tools can be applied across almost any industry and there are many off-the-peg, as-a-service solutions that can be deployed in areas such as customer service, content generation, and business reporting, without requiring huge investment. As the technology improves and the practical uses widen, expect NLP and NLG to become even more influential in the near future.

That's not to say you should rush into adopting NLP and NLG. As with most of the trends in this book, to get the most out of the technology, you must approach it in a strategic way. What I mean by that is there must be a strategic reason to introduce any new technology – and that reason may be helping to achieve a specific business objective or solve a certain problem.

Whatever tool or application you choose, remember that the technology is developing rapidly, which means you'll need to be prepared to adapt and learn as you go along. This is par for the course with any AI-based technology, since a critical facet of AI is machines' ability to learn from data and get better over time.

Notes

1. Machines Beat Humans on a Reading Test. But Do They Understand? *Quanta Magazine*: www.quantamagazine.org/machines-beat-humans-on-a-reading-test-but-do-they-understand-20191017/
2. This man quit his job and built a whole company so he could talk to his daughter: www.weforum.org/agenda/2018/01/this-man-made-an-app-so-he-could-give-his-daughter-a-voice/
3. SignAll: www.signall.us/
4. Meet 4 Grammarly Users Who Will Inspire You: www.grammarly.com/blog/meet-inspiring-grammarly-users/
5. Lenddo: https://lenddo.com/
6. Swedish Bank Uses Natural Language Processing for Virtual Customer Assistance: https://emerj.com/ai-case-studies/swedish-bank-uses-natural-language-processing-virtual-customer-assistance/
7. Textio Hire: https://textio.com/products/
8. Emerging federal use cases: www.accenture.com/us-en/insights/us-federal-government/nlp-emerging-uses

9. The Rise of the Robot Reporter, *New York Times*: www.nytimes.com/2019/02/05/business/media/artificial-intelligence-journalism-robots.html

10. Entering The Next Century With A New Forbes Experience, *Forbes*: www.forbes.com/sites/forbesproductgroup/2018/07/11/entering-the-next-century-with-a-new-forbes-experience/#6b49d3b3bf4f

11. The Washington Post's robot reporter has published 850 articles in the past year: https://digiday.com/media/washington-posts-robot-reporter-published-500-articles-last-year/

12. Automated identification and predictive tools to help identify high-risk heart failure patients: pilot evaluation: www.ncbi.nlm.nih.gov/pubmed/26911827

13. Natural language processing in healthcare: www.mckinsey.com/industries/healthcare-systems-and-services/our-insights/natural-language-processing-in-healthcare

14. Dragon Medical One: www.nuance.com/en-gb/healthcare/physician-and-clinical-speech/dragon-medical-one.html

TREND 11
VOICE INTERFACES AND CHATBOTS

The One-Sentence Definition

Voice interfaces and chatbots are computer programs that allow humans to converse and interact with computers through either spoken commands or written text.

What Are Voice Interfaces and Chatbots?

Computer programs that simulate human conversation have, within the space of just a few years, rocketed into everyday use. Both voice interfaces and chatbots work in a similar way, using artificial intelligence (AI) and deep learning (Trend 1), Big Data (Trend 4), and natural language processing and natural language generation (Trend 10) to both understand and respond to human speech. Although they're underpinned by the same technologies, voice interfaces (including digital assistants like Siri and smart speakers like Amazon's Alexa) and chatbots interact with users in slightly different ways. Voice interfaces respond to spoken language commands (which is really useful for those languages that are easier to speak than type – like Chinese – or when the user is unable to type), while chatbots interact with people through a written chat interface, such as Facebook Messenger or a web-based application. In both cases, the computer uses natural language processing to understand the text, and then analyzes the text using AI and deep learning algorithms to determine the best response.

Voice interface tools in particular have proven massively popular with consumers, particularly when it comes to smart speakers. In the United States in 2018, ownership of smart speakers rose 39.8% to 66.4 million, with the Amazon Echo (featuring Alexa) being the clear market leader.[1] The same report also found that smart speaker ownership is driving increased usage of voice assistants on smart phones.

The technology has in fact been around for decades – the first "chatterbot" Eliza was developed in 1994[2] – but the massive leaps in AI and deep learning over the last five years or so have dramatically improved the ability of chatbots and voice assistants to converse with humans more naturally. While early conversations with Eliza were pretty basic, today's voice interface and chatbot technology is so impressive that it's not always possible to tell whether you're interacting with a bot or a human. What's more, the technology is getting better all the time – advancing at an incredible pace – to the extent where machines can do a lot more than simply understand our speech (this in itself is no mean feat, when you think about how nonlinear our spoken conversations tend to be, with lots of interruptions, repetition, pauses, slang, and words with multiple meanings). Today, the technology is so advanced that computers can now understand the nuances of human emotion and even detect whether you're lying.

For example, the Woebot chatbot therapist interprets data on a person's mood and helps them discuss their mental health – the thinking being that people are more likely to open up to a bot because they know they won't be judged.[3] And when it comes to detecting lies, researchers at Florida State University and Stanford say they've developed the first online polygraph system – which can separate lies from truth without the need for face-to-face monitoring.[4] Interestingly, research also suggests that people are more likely to be honest with a robot; a screening system built by the National Center for Credibility Assessment in the US found that candidates were much more likely to admit to experiencing mental health issues, using illicit substances, or committing crimes to the on-screen avatar than in a written questionnaire.[5]

So, while Alexa, Siri, and Cortana are perhaps the best-known examples of this technology in action, today's intelligent bots are capable of so much more than just telling you the weather forecast or playing your children's favorite song for the thousandth time. In fact, they're already impacting the way we live and how organizations interact with their customers.

How Are Voice Interfaces and Chatbots Used in Practice?

In the world of business, voice interfaces and particularly chatbots are (for now) predominately being used in customer service, marketing, and sales functions – but there are examples across lots of business functions and industries.

Let's look at some real-world use cases that demonstrate the benefits of voice interfaces and chatbots for organizations:

- UK retailer **Marks & Spencer** added a virtual digital assistant function to its website to help customers troubleshoot discount codes and other common issues without needing human intervention. The company claims the bot has saved online sales worth £2 million that it would otherwise have lost.[6]

- **Whole Foods** has a Facebook Messenger chatbot that serves up recipes and cooking inspiration, thereby deepening customers' relationship with the brand. The bot can understand emojis as well as text.

- **Asos** was able to triple orders and reach 35% more people using Messenger chatbots.[7]

- Popular European food retailer **Lidl** has created a wine bot called Margot to help customers get the most out of its extensive wine range. Margot chats with customers via Facebook Messenger to give them tips on wine-and-food pairings and tell them more about the winemaking process.

- **UNICEF** uses a chatbot to conduct surveys and gather data around the world. Information gathered by the U-Report platform can have a real impact on the organization's policy recommendations – for example, one poll of Liberian children found that 86% of respondents said their schools had a problem with teachers giving children better grades in return for sex, prompting action by the Liberian Minister of Education. Thanks to fast, cost-effective chatbot technology, UNICEF was able to poll 13,000 Liberian children in just 24 hours.[8]

- Travel company Hipmunk has a digital virtual assistant (called **Hello Hipmunk**) that helps users book flights, hotels, and car rentals to plan the perfect trip without talking to a human travel agent.

- Philippines-based **Globe Telecom** decreased call volume by 50% and increased customer satisfaction by 22% using a Facebook Messenger chatbot. Employees were 3.5 times more productive as a result of the new system.[9]

- Chatbot **Polly** is designed to improve workplace happiness by conducting surveys and gathering employee feedback, allowing organizations to keep track of how employees are feeling about the workplace and nip morale problems in the bud before they escalate.

- The **Voca** voice interface system allows companies to reach out to customers and potential customers on a large scale with a personal, human voice – that's generated by a computer. Voca can fulfill a number of tasks, such as making those boring, repetitive cold calls for you, and passing promising leads onto human sales reps. This frees up the sales force to work on the most valuable leads only, and could revolutionize work for anyone who hates making cold calls. Leading management consultancy firm McKinsey estimates that 36% of sales rep work could be automated by bots.[10]

- **ROSS Intelligence** is an AI-driven research assistant that can perform legal research for law firms. Users reduced their research time by more than 30%.[11]

- The **US Army** is using a chatbot named SGT STAR to quickly answer questions about joining the service and help enlist the soldiers of the future.

Outside of work, bots are being used in all sorts of creative ways to improve our everyday lives:

- **Marvel** has a chatbot that gives fans the chance to chat with Spiderman.

- The **HealthTap** chatbot responds to medical questions, concerns, and symptoms. And if the chatbot is stumped by a user's query, it'll be referred to a human health professional for answers.

- **Insomnobot 3000** keeps insomniacs company while the rest of the world sleeps.

- The **Endurance** chatbot has friendly conversations with patients who have suspected Alzheimer's and other forms of dementia to test their ability to remember information. This can help with diagnoses and tracking memory recall over time.

- **Vi** is a digital fitness coach and personal trainer that gets to know you and your fitness goals, and provides personalized workouts.

- **Google Duplex** is one of my favorite examples of a digital assistant, and it's well worth watching a video or listening to an audio clip of the system in action. Using voice interface technology, Google Duplex does the talking for you, calling your hairdresser, dentist, local restaurant, or whomever to make appointments and enquiries for you. The voice interface is absolutely uncanny, and seamlessly responds to the person on the other end of the line in a way that's unbelievably authentic. It even throws in all the little "ums" and "ahs" that litter our everyday speech.[12]

Chatbot and voice technology is now becoming so good that the line between professional assistant and social companion or friend is becoming blurred. As an example, Microsoft's chatbot Xiaoice has proven a massive hit in China, attracting 660 million users. In fact, Xiaoice is so wildly popular, she ranks as one of China's most admired celebrities, and receives love letters and gifts from adoring fans.[13] The secret of her success lies in the fact that Xiaoice is gradually learning to interact with humans using social skills, nuance, and emotions. Some users spend hours chatting with her.

Elsewhere, New York-based startup Hugging Face wants its social AI to be your teenage child's new BFF. It chit-chats and trades selfies, and is proving a big hit with young users – the app receives over 1 million messages a day.[14]

Replika is another example of an artificial companion that won't help you book a table at your favorite restaurant – but it will chat with you for hours. Thanks to deep learning, over time, Replika learns to speak like the person it's conversing with.[15]

Key Challenges

Exciting as this technology is, there are some challenges – ethical, practical, and technological – that you'll need to consider if you want to incorporate voice interfaces or chatbots into your business.

Let's start with the ethical. Check out a video or audio clip of Google Duplex in action and it seems like the person on the other end of the phone has no idea they're talking to a machine. This throws up an ethical dilemma – is it okay to let people think they're conversing with a real-life human? In my view, it's not okay. Ideally, you should make it clear to those on the other end of the conversation that they're actually interacting with a computer.

Also, there are times when voice interfaces and chatbots just aren't appropriate. Yes, the technology has advanced dramatically in recent years and bots are now capable of holding incredibly natural conversations, and even understanding human emotion – but there are times when only human interaction will do. To avoid alienating your target audience, it's vital you think about which tasks are best suited to bots and which should be reserved for human responders. For example, a busy HR department may use chatbots to answer simple employee questions like "How many days' holiday do I have left?" but what if an employee wants to lodge a grievance or get advice on harassment? That's definitely the realm of a human advisor. Essentially, organizations need to find the sweet spot between when bots are appropriate and when the human touch is needed. And sometimes users would simply prefer to talk to a human, even for a quick and simple issue. Not giving users the option of bypassing the chatbot or voice interface could be a mistake.

From a practical point of view, one error that many businesses make is to confuse the notion of voice interfaces and chatbots with self-service offerings like online FAQ pages or voice menus on a phone system. If your customers have access to tools and information that helps them solve their own problems, and that works for your customers, that's great. But let's be clear: voice interfaces and chatbots offer more than just helping customers solve a particular problem or complete a certain task; these tools provide a genuine sense of interacting with a company – by engaging the customer in a natural conversation, the interaction is much more meaningful.

Another mistake is not keeping the target audience in mind when creating your chatbot or voice interface. The end user must be at the heart of every decision you make – from the means of communication (for example, do you want to communicate via Facebook Messenger?) to the language and tone of voice used.

What's more, as with most of the trends in this book, it's about incorporating the technology in a strategic way – in a way that adds real value. There must be a reason for introducing a voice interface or chatbot in the first place, whether it's to drive sales, improve the customer experience, answer queries more quickly, provide a more personalized customer experience, or whatever. Adopting new technology for the sake of it is never a good idea.

Finally, limitations in the technology itself can result in an underwhelming user experience. For example, if the bot doesn't seamlessly understand users' language, it'll quickly frustrate and alienate your customers. Or even if the system understands perfectly, if the responses seem flat and robotic, that will also put people off. The customer might go away with the answer they were looking for, but lack that warm feeling of having a positive, meaningful interaction with a brand. This will become less of an issue as the technology continues to evolve, but be prepared to tweak and improve your offering in line with user feedback.

How to Prepare for This Trend

There are lots of user-friendly, accessible, and affordable tools on the market that make it possible for any business to create a chatbot or voice interface system. These can deliver really powerful ways to automate and improve customer service transactions, sales and marketing interactions, and much more – providing your customers with 24-hour-a-day access, potentially in a variety of languages.

So the good news is, you don't need to be a tech guru to make use of this technology. Many of the tools are available as-a-service, meaning you don't need to invest in lots of new infrastructure or in-house expertise. However, because these tools are developing so fast, you'll definitely need to stay up to date on what's possible and be prepared to tweak and expand your offering as new possibilities arise.

If you're keen to use chatbots or voice interfaces in your business, I recommend you follow these general steps and practical tips:

- **Identify your goals and audience.** To get the most out of this technology, you must be absolutely clear on what problems you want to solve, what improvements you want to deliver, what processes you want to streamline, and who you want to help. Think about both your internal business needs here and the needs of your customers or users.

- **Check out your competitors.** Look at how your competitors are using chatbots and voice interfaces – not to copy them, but to see what you might do better. To inform this step, think about how you already differ from your competitors, and how that might influence your use of bots.

- **Think about the personality you want to project.** After all, interactions via a chatbot or voice interface must be consistent with the rest of your brand, your typical communication style, and the kind of interaction your customers have come to expect.

- **Choose a platform and provider.** Thankfully, there are lots of available tools that don't require any in-depth technical expertise, such as Chatfuel, Flow XO, and Voca. Many will provide a free trial so you can see how the technology works and what benefits it could deliver to your business.

- **Don't be afraid to experiment.** Things move fast these days and if you wait for a system to be 100% perfect before launching, chances are it'll be out of date before you manage to roll it out. So start quickly – and remember, you can always start small in one area of the business, gather user feedback, and expand or improve your offering from there.

- **Be prepared to adapt and learn as you go.** The thing about AI-based tools is that they get smarter as they go along, so expect your bots to be continuously learning and getting better. In other words, rather than this being a service you launch and forget about, you'll almost certainly need to make regular tweaks and improvements.

- **Measure success.** Voice interfaces and chatbots must deliver value for your business. Think about how you'll measure the return on your investment and make certain you're getting the desired results.

Notes

1. U.S. Smart Speaker Ownership Rises 40% in 2018 to 66.4 Million and Amazon Echo Maintains Market Share Lead Says New Report From Voicebot: https://voicebot.ai/2019/03/07/u-s-smart-speaker-ownership-rises-40-in-2018-to-66-4-million-and-amazon-echo-maintains-market-share-lead-says-new-report-from-voicebot/
2. A brief history of Chatbots: https://chatbotslife.com/a-brief-history-of-chatbots-d5a8689cf52f?gi=74afa943f773
3. How Chatbots Are Learning Emotions Using Deep Learning, *Chatbots Magazine*: https://chatbotsmagazine.com/how-chatbots-are-learning-emotions-using-deep-learning-23e1085e4cfe
4. Researchers Built an "Online Lie Detector." Honestly, That Could Be a Problem, *Wired*: www.wired.com/story/online-lie-detector-test-machine-learning/
5. US government chatbot gets you to tell all, *New Scientist*: www.newscientist.com/article/dn25951-us-government-chatbot-gets-you-to-tell-all/
6. A year in, Marks & Spencer's virtual assistant has helped drive £2 million in sales: https://digiday.com/marketing/year-marks-spencers-virtual-assistant-helped-drive-2-5m-sales/

7. Fueling growth through mobile: www.facebook.com/business/success/asos

8. Success Story: U-Report Liberia exposes Sex 4 Grades in school: https://ureport.in/story/194/

9. Building customer relationships with Messenger: www.facebook.com/business/success/globe-telecom

10. Chatbot Report 2018, *Chatbots Magazine*: https://chatbotsmagazine.com/chatbot-report-2018-global-trends-and-analysis-4d8bbe4d924b

11. ROSS AI Plus Wexis Outperforms Either Westlaw or LexisNexis Alone, Study Finds: www.lawsitesblog.com/2017/01/ross-artificial-intelligence-outperforms-westlaw-lexisnexis-study-finds.html

12. Google Duplex rolling out to non-Pixel, iOS devices in the US: https://9to5google.com/2019/04/03/google-duplex/

13. Much more than a chatbot: China's Xiaoice mixes AI with emotions and wins over millions of fans: https://news.microsoft.com/apac/features/much-more-than-a-chatbot-chinas-xiaoice-mixes-ai-with-emotions-and-wins-over-millions-of-fans/

14. Hugging Face's artificial intelligence wants to become your artificial BFF: www.prnewswire.com/news-releases/hugging-face-s-artificial-intelligence-wants-to-become-your-artificial-bff-828267998.html

15. The emotional chatbots are here to probe our feelings, *Wired*: www.wired.com/story/replika-open-source/

TREND 12
COMPUTER VISION AND FACIAL RECOGNITION

The One-Sentence Definition

Computer vision, also referred to as machine vision, is where machines (including computers, software, and algorithms) can "see" and interpret the world around them – with facial recognition (which uses computer vision to identify people) being a prime example.

What Is Computer Vision and Facial Recognition?

Early experiments in computer vision began as far back as the 1950s, and the technology was already being used commercially by the 1970s to interpret typed and handwritten text.[1] So, if it's not a new technology, why highlight it as a key trend today? To answer that question, we first need to get a quick (and non-technical) explanation of how computer vision works.

As a form of artificial intelligence (AI, see Trend 1), computer vision is essentially all about processing, analyzing, and making sense of data – it's just that the data being analyzed is visual rather than, say, textual or numerical. For the most part, this means the data being analyzed is in the form of photos or videos, but it could also include data from thermal and infrared cameras and other visual sources.

Analyzing visual data with any real accuracy relies on deep learning and neural networks (Trend 1) – in other words, using pattern recognition to distinguish what's in an image, after learning from a specific data set of other relevant images. For example, in 2012, Google was using a neural network to identify cat videos on YouTube (a worthy application if ever there was one). To learn to recognize cats, the system needed lots of images, some containing cats and some without cats. Crucially, because of deep learning – which means the computer system learns to train itself – programmers didn't have to tell the system what signifies a cat (i.e. whiskers, tail). Instead, the system learned for itself by trawling through millions of images.

This shows how, like any form of AI, computer vision is absolutely reliant on data ... and lots of it. And that's why computer vision has exploded into everyday use in recent years: we're now generating more data than ever before (see big data, Trend 4), and much of this data is visual. Ninety-five million photos and videos are shared each day on Instagram alone.[2] And don't forget all the non-Instagram-worthy snaps we take every day, or the use of CCTV security cameras around the world.

The sheer volume of data we're generating every day is a major driver in the growth of computer vision. As well as this, computing power has also evolved to make it easier and cheaper to store and process hefty image data. These two factors have combined to rapidly make computer vision more common, more accessible, and more accurate – in less than a decade, computer vision progressed from 50% accuracy to 99%, making computers more accurate than humans at quickly reacting to visual data.[3] As the technology is becoming increasingly cheaper and easier to deploy, it's no wonder the entire machine vision market is predicted to reach $14 billion by 2024 (up from $9.9 billion in 2019).[4]

Facial recognition is a subset of computer vision. Like your finger-print, your faceprint is a unique code that's applicable to you; but unlike your fingerprint, your faceprint can be scanned at a distance, without you even realizing it's being done. As we'll see in some of the practical examples that follow, facial recognition technology is being used more widely today than you might realize, particularly in China.

How Is Computer Vision and Facial Recognition Used in Practice?

Computer vision and facial recognition is being used in settings as diverse as manufacturing, healthcare, autonomous vehicles, and secu-rity and defense. The technology is such a part of everyday life nowa-days that you very likely experience it regularly without necessarily recognizing it.

To begin, let's look at some of my favorite examples of computer vision and facial recognition at work in everyday life:

- Scratching your head at a foreign sign or menu while on holi-day? Using **Google Translate**, all you have to do is point your phone's camera at the words and, hey presto, Google will trans-late it into your preferred language almost instantly – all thanks to computer vision. The app uses a process called optical char-acter recognition to "see" the text; then uses augmented reality (AR, see Trend 8) to overlay the translation over the top of the original text.

- In the **healthcare** sector, a staggering 90% of all medical data is image based,[5] meaning there are many valuable uses for com-puter vision. Microsoft's InnerEye software is one such exam-ple. The system can analyze X-ray images and identify possible tumors and other anomalies – it then automatically flags these areas for further analysis by a human radiologist. The system has been used by Addenbrooke's hospital in Cambridge, England,

and been highlighted by the government as an example of how AI could help to transform the UK's National Health Service.[6]

- When it comes to facial recognition technology, **China** is racing forward to become a world leader.[7] The Beijing Subway is planning to use facial recognition systems in place of tickets and, on the streets of the same city, police officers wear AR glasses that are able to cross-reference faces against the national database to spot criminals. Chinese police have also used facial recognition technology to find four missing children.[8]

Now let's focus on the world of business, where computer vision is finding many valuable uses across a wide range of industries.

- Computer vision is partly what enables **autonomous vehicles** – like those being produced by Tesla, BMW, and Volvo – to safely drive on roads, navigate around objects, change lanes, and "see" road signs and traffic signals, using multiple cameras and sensors to interpret what's going on around them and respond accordingly. Particularly in the transport and logistics industry, companies are bracing for the impact of autonomous trucks – although we're likely several years away from fully autonomous lorries that need no driver involvement across the entire journey. In the immediate future, autonomous developments in trucks will focus on "platooning" (where a convey of trucks travels together on motorways with a driver only in the lead vehicle), or with human drivers taking over for the more complex pickup and drop-off stages of the journey.[9]

- In agriculture, **John Deere's** semi-autonomous combine harvester uses AI and computer vision to find the optimal route through a field of crops, and to analyze the quality of grains as they're being harvested. The company also hopes computer vision will help farmers cut the amount of herbicides needed by 90% – because machinery in the field will be able to use computer vision to tell the difference between healthy crops, which don't need spraying, and unhealthy crops.[10] Elsewhere in agriculture, computer vision is being used to detect the ripeness of papayas,[11] and to sort and classify cucumbers.[12]

- **Shanghai Airport** has introduced an automated clearance system that uses facial recognition. Passengers scan their ID cards and use security-checking machines that are equipped with facial recognition technology to complete the security check process – all in 12 seconds.[13]

- Two **Marriott hotels** are using facial recognition to speed up the check-in process. Guests at Hangzhou Marriott Hotel Qianjiang and Sanya Marriott Hotel Dadonghai Bay, both in China, can check in using kiosks equipped with facial recognition technology. After guests scan their ID, the system takes a photo and confirms their identity; then the kiosk dispenses their room key. Similarly, **Royal Caribbean Cruises** is using facial recognition technology to speed up the boarding process – plus, onboard, computer vision is used to detect congestion as passengers move around the ship.[14]

- **Disney** also uses computer vision to enhance the customer experience. Disney Research is tracking the reactions of audiences to films using computer vision. Cameras monitor test audiences at film previews and the data can be analyzed to gauge audience sentiment.[15] In the future, this technology could be introduced into other Disney experiences, such as its amusement parks.

- **Walmart** is using computer vision at checkouts in more than 1,000 stores to combat "shrinkage" – loss from theft and scanning errors, to you and me. In an initiative known as Missed Scan Detection, cameras are used at both self-checkout machines and regular, cashier-manned checkouts to automatically identify when an item hasn't been scanned properly (whether by accident or on purpose). When an issue is identified, the system alerts a member of staff so they can step in. With estimates suggesting shrinkage at Walmart could equate to more than $4 billion a year, technology like this could potentially have a huge impact on the bottom line.[16] So far, shrinkage rates have declined at stores where Missed Scan Detection has been deployed.

- Elsewhere in retail, Amazon is eliminating the checkout process altogether in its small, but growing, chain of **Amazon Go** grocery and convenience stores.[17] The customer simply scans themselves in (using the Amazon app on their smart phone) at a turnstile when they enter the shop, picks up what they want from the shelves, then leaves – no queuing at checkouts, no handing over cash, no "unexpected item in the bagging area." Cameras track you as you shop, monitor what you take, and the cost is charged to your Amazon account automatically.

- Computer vision is also being used to detect when faces in photos have been manipulated or photoshopped. Researchers at **Adobe and UC Berkeley,** who have teamed up on the project, hope it will combat the rise of deepfakes. And it looks like they're onto a winner, as tests showed the tool was 99% accurate at detecting altered images, and it could even revert images to what it predicted was their undoctored state.[18] Tools like this could prove immensely useful for the media.

- In **manufacturing settings**, machine vision can be used for predictive maintenance (basically, predicting and fixing problems before they occur), health and safety, quality control, and more. FANUC's Zero Down Time solution collects and analyzes images from manufacturing equipment to identify signs that components might malfunction, so that they can be replaced or repaired before a failure occurs – thereby reducing costly downtime. In an 18-month pilot, the tool was tested across 38 automotive factories where it was able to detect and prevent 72 failures.[19]

- And in food production, pizza giant **Domino's** is using computer vision at over 2,000 locations to ensure its pizzas are top quality.[20] The "Pizza Checker" camera system can even distinguish between different types of pizza and confirm the pizza is at the right temperature. Results are sent to a store manager, and a photo can be sent to the customer. Customers will also be notified if their pizza failed its quality control test and has to be redone.

- While we're on the subject of fast food, one branch of **KFC** in Hangzhou, China, has been testing a payment system that analyzes your smile to confirm your identity and take payment (via the Alipay app) instead of paying with cash or card – an exciting development that could drastically help to cut fraud in future.[21]

- Chinese facial recognition technology **Megvii** is known for its Face++ technology, the system that underpins the Chinese law enforcement and "smile to pay" KFC examples. At Megvii's own offices, personnel don't access the building using normal security badges or passes – it takes a flash of a smile, which is then analyzed against the company's personnel database.[22] The same technology could be used in any security-conscious building, whether commercial or residential, to make sure only authorized people are allowed in.

- Sticking with security, **Evolv Technology** offers a physical security system that can screen up to 900 people an hour using facial recognition technology, thereby eliminating bottlenecks and queues at busy events.[23] Evolv says its system can be programmed with photos of VIPs, season ticket holders, priority customers, and people you don't want gaining entry. (If a "person of interest" is spotted, the system can either block their entry, or flag their identity for a human security officer to verify.) The scanning machine can be moved around to create checkpoints wherever they're needed.

Key Challenges

One of the biggest challenges, particularly around facial recognition, is the issue of privacy – and in the Western world there have been a number of examples of individuals and campaign groups challenging the use of facial recognition in public. For example, in the UK, office worker Ed Bridges has brought a case against South Wales Police, claiming they violated his privacy and data protection rights when they used facial recognition technology on him (Mr Bridges says his face was scanned while shopping in 2017 and during a peaceful protest in 2018).[24] At the time of writing, the case was still under consideration, but it may have far-reaching implications for the use of facial recognition software in the UK. Many argue that use of the technology is unregulated, although the police have responded that they comply with data protection rules.

Cases like this indicate a general unease among people about the idea of being "spied on" as they go about their daily, law-abiding business – plus a sense that the technology is advancing at such a rate, legislation and best practice guidelines just can't keep up. The surveillance commissioner for England and Wales, Tony Porter, has openly said the code of practice for the use of surveillance cameras needs to be strengthened.[25] And one government advisory group, the Biometrics and Forensics Ethics Group, has said facial recognition should only be used in law enforcement it if is proven to be effective at identifying people, can be used without bias, and when there's no other method available.[26]

Given all this, we may see legislation introduced in some countries and areas to restrict or oversee the use of facial recognition software. San Francisco is leading the way in this and has already banned the use of facial recognition technology by police and other agencies.[27] Also in the US, shareholders have been trying to prevent Amazon selling its facial recognition software to police forces – although they were defeated in votes at the company's annual general meeting.[28] Clearly, any business planning to use facial recognition must keep abreast of these developments in the ethical use of the technology.

How to Prepare for This Trend

If there's one message I want to leave you with in this chapter it's to remember that AI is fantastic at pattern recognition – and because of this, many business processes can be automated and improved with computer vision. In any part of your business that generates or has the potential to generate visual data, harnessing AI's talent for pattern recognition could pay dividends. I'd therefore advise any business to think about their unique challenges and bottlenecks to see whether computer vision could help smooth out and improve those processes.

Notes

1. Computer Vision: What it is and why it matters: www.sas.com/en_us/insights/analytics/computer-vision.html
2. 33 Mind-Boggling Instagram Stats & Facts for 2018: www.wordstream.com/blog/ws/2017/04/20/instagram-statistics
3. Computer Vision: What it is and why it matters: www.sas.com/en_us/insights/analytics/computer-vision.html
4. $14 Bn Machine Vision Market: www.businesswire.com/news/home/20190528005387/en/14-Bn-Machine-Vision-Market—Global
5. IBM Watson Health, Merge launch new personalized imaging tools at RSNA: www.healthcareitnews.com/news/ibm-watson-health-merge-launch-new-personalized-imaging-tools-rsna
6. Project InnerEye – Medical Imaging AI to Empower Clinicians: www.microsoft.com/en-us/research/project/medical-image-analysis/
7. The Fascinating Ways Facial Recognition AIs Are Used in China, Bernard Marr: www.forbes.com/sites/bernardmarr/2018/12/17/the-amazing-ways-facial-recognition-ais-are-used-in-china/#3700d91f5fa5
8. Chinese police track four missing children using AI, *People's Daily Online*: http://en.people.cn/n3/2019/0619/c90000-9589632.html
9. Distraction or disruption? Autonomous trucks gain ground in US logistics: www.mckinsey.com/industries/travel-transport-and-logistics/our-insights/distraction-or-disruption-autonomous-trucks-gain-ground-in-us-logistics

10. Blue River See & Spray Tech Reduces Herbicide Use By 90%, AG Web: https://www.agprofessional.com/article/blue-river-see-spray-tech-reduces-herbicide-use-90

11. AI Detects Papaya Ripeness: https://spectrum.ieee.org/tech-talk/robotics/artificial-intelligence/ai-detects-papaya-ripeness

12. How a Japanese cucumber farmer is using deep learning and TensorFlow: https://cloud.google.com/blog/products/gcp/how-a-japanese-cucumber-farmer-is-using-deep-learning-and-tensorflow

13. Shanghai airport first to launch automated clearance system using facial recognition technology, *South China Morning Post*: www.scmp.com/tech/enterprises/article/2168681/shanghai-airport-first-launch-automated-clearance-system-using

14. AI on Cruise Ships: www.bernardmarr.com/default.asp?contentID=1876

15. Disney Uses Big Data, IoT And Machine Learning To Boost Customer Experience, *Forbes*: www.forbes.com/sites/bernardmarr/2017/08/24/disney-uses-big-data-iot-and-machine-learning-to-boost-customer-experience/#123a1b233876

16. Walmart reveals it's tracking checkout theft with AI-powered cameras in 1,000 stores: www.businessinsider.com/walmart-tracks-theft-with-computer-vision-1000-stores-2019-6?r=US&IR=T-

17. Computer Vision Case Study: Amazon Go. *Medium*: https://medium.com/arren-alexander/computer-vision-case-study-amazon-go-db2c9450ad18

18. Adobe trained AI to detect facial manipulation in Photoshop: www.engadget.com/2019/06/14/adobe-ai-manipulated-images-faces-photoshop/

19. 10 Examples of Using Machine Vision in Manufacturing: www.devteam.space/blog/10-examples-of-using-machine-vision-in-manufacturing/

20. Domino's Will Use AI to Make Sure Every Pizza They Serve is Perfect: https://interestingengineering.com/dominos-will-use-ai-to-make-sure-every-pizza-they-serve-is-perfect

21. The Fascinating Ways Facial Recognition AIs Are Used in China, *Forbes*: www.forbes.com/sites/bernardmarr/2018/12/17/the-amazing-ways-facial-recognition-ais-are-used-in-china/#3700d91f5fa5

22. The Amazing Ways Chinese Face Recognition Company Megvii (Face++) Uses AI and Machine Learning, *Forbes*: www.forbes.com/sites/bernardmarr/2019/05/24/the-amazing-ways-chinese-face-recognition-company-megvii-face-uses-ai-and-machine-vision/#5291b5e312c3

23. AI for Physical Security: 4 Current Applications: https://emerj.com/ai-sector-overviews/ai-for-physical-security/

24. Facial recognition tech prevents crime, police tell UK privacy case, *The Guardian*: www.theguardian.com/technology/2019/may/22/facial-recognition-prevents-crime-police-tell-uk-privacy-case

25. Surveillance camera czar calls for stronger UK code of practice, *Computer Weekly*: www.computerweekly.com/news/252465491/Surveillance-camera-czar-calls-for-stronger-UK-code-of-practice

26. Cops told live facial recog needs oversight, rigorous trial design, total protection against bias, *The Register*: www.theregister.co.uk/2019/02/27/biometrics_forensics_ethics_facial_recognition/

27. San Francisco Bans Facial Recognition Technology, *New York Times*: www.nytimes.com/2019/05/14/us/facial-recognition-ban-san-francisco.html

28. Amazon heads off facial recognition rebellion: www.bbc.com/news/technology-48339142

TREND 13
ROBOTS AND COBOTS

The One-Sentence Definition

Today's robots can be defined as intelligent machines that can understand and respond to their environment and perform routine or complex tasks autonomously.

What Are Robots and Cobots?

In this data-driven age, it's the intelligence and ability to act autonomously that define robots and separate them from other machines.

We've had equipment that can automate industrial functions for hundreds of years, but, surprisingly, the word "robot" wasn't coined until 1920. Czech writer Karel Capek, best known for his science fiction, used the word in his play R.U.R. (translated as Rossom's Universal Robots) to describe artificial automata. (In the play, the robots ended up going on a killing spree, which might explain the origins of our mistrust of robots.)

The first industrial robot, called Unimate, was invented in 1950. Early industrial robots would be programmed to complete certain functions in, say, manufacturing settings, and were used to replace repetitive manual labor. In the last couple of decades, robots have become more advanced, with greater intelligence and more automation. This is largely thanks to artificial intelligence (AI) (Trend 1), sensors and the IoT (Trend 2), and big data (Trend 4). Without the advances in these fields, many of the amazing examples I outline later in the chapter wouldn't be possible. Today's robots are not only physically more robust and flexible than early industrial robots, they're also much smarter. We have delivery robots, robots that can perform surgery, robots for space exploration, demolition robots, underwater robots, search-and-rescue robots, and more. We have robots that can walk, run, roll, jump, and even backflip.[1]

Robots are now commonplace in sectors like car manufacturing (the International Federation of Robotics has forecasted that 1.7 million new robots will be installed in factories around the world by 2020)[2] but they're starting to make their move into other sectors. According to one estimate, up to 35% of organizations in health, utilities, and logistics are exploring the use of automated robots.

Robots are also making their way into our homes. Those cute vacuum cleaners that look like hockey pucks are perhaps the most widely known example of robots in the home. But, in the future, might we see robots take on a wider variety of tasks in the home, like keeping elderly people company or caring for pets when their owners are at work? How popular these domestic robots would be remains to be seen, but technology company Nvidia is betting robots for the home will become a commercial success; the company is partnering with Ikea to develop a robotic kitchen assistant.[3]

Then we have the rise of collaborative robots, or *cobots*. This latest generation of robotic systems is designed to work alongside humans – as robotic colleagues – helping to enhance the work that humans do, and interacting safely and easily with the human workforce.

Think of cobots as extra robotic muscle in the workplace. Thanks to AI technologies like machine vision (Trend 12), cobots are able to sense the humans around them and react accordingly – for example, by adjusting their speed or reversing to avoid humans and other obstacles. This means workflows can be designed to get the very best out of both humans and robots working together. An Amazon fulfillment center is a great example of this in action, with robots bringing items to human workers for packing. And with an average price tag of around $24,000 each,[4] cobots look set to be a viable option to help smaller and mid-sized firms compete with larger organizations.

Cobots help to put a friendly face on automation. Advances in robotics and AI have left many worried about losing their jobs to a machine. To a certain extent, automation is an inevitable worry for every generation of workers, but cobots show that the future of work (at least in the short and medium term) is likely to involve working alongside a robot, rather than being replaced by one.

Ultimately, I believe robots will help to free up humans from the 4Ds: that is, work that is *dull, dirty, dangerous and dear (expensive)*:

- Dull work that is repetitive and tedious, leaving humans to focus on more creative or rewarding tasks.

- Dirty jobs that keep our world functioning, but that most of us don't give a second thought to, such as sewer reconnaissance.

- Dangerous jobs, such as investigating and detonating bombs.

- Dear or expensive jobs, where using a robot saves money or reduces delays.

There are also interesting advances in the development of humanoid robots that look increasingly lifelike (an effect known as "Uncanny Valley"). However, the robotics field appears to be split on whether robots should actually look like us. On the one hand, some think humans will find it easier to interact with robots that look more like us. Others, however, find the idea distinctly creepy.

Who knows what's in store for the future of robotics, but one thing is certain: the robots are here to stay.

How Are Robots and Cobots Used in Practice?

Let's look at some exciting (and a few weird) real-life examples of robots in a variety of settings.

Delivery Robots

Next time you open your door for a delivery, it could be a robot delivering your parcel or takeaway. Robot delivery devices are hotly tipped to solve the "last mile" problem of delivery operations – the final, most expensive stage of the delivery process.

- **Starship Technology's** autonomous delivery robots are already on the streets in my hometown of Milton Keynes. I use them to deliver groceries from the local Co-op, but they're also being used to deliver, among other things, Just East takeaways in London, Domino's pizzas in Hamburg, Germany, and food to students on university campuses across the US. Looking a bit like a minifridge on wheels, the Starship robots have a maximum speed of 10 miles per hour, have clocked up an impressive 350,000 miles, and are hugely efficient in cities.[5]

- A little larger than the Starship robot is **Nuro**. Designed by a team of Google engineers, and around half the size of a small car, Nuro robots have been delivering groceries in Arizona and Texas. In 2019, the company announced it was branching out into pizza delivery.[6]

Robots for Safety and Security

Robots are also being used to help keep us safe, or perform work that's dangerous for humans.

- **Cobalt Robotics** offers a robot security platform that it says is 65% cheaper than human security guards.[7]

- **Dubai** has unveiled a robotic police officer to patrol city malls and tourist attractions. The aim is for 25% of Dubai's police force to be robotic by 2030.[8]

- The **GoBetween** robot is designed to make traffic stops safer, by sliding out of the front of the police car and acting as the first point of contact with the driver in front. The robot has a camera, speaker, and microphone, and can print traffic tickets from its chest.[9]

- **Colossus** is a firefighting robot that helped battle the devastating 2019 fire at Notre-Dame in Paris.[10] It's like WALL-E, but with a water cannon.

You might also like to check out autonomous military drones in Trend 19.

Robots in Healthcare

Robots are slowly beginning to change the face of medicine and augment the work of human healthcare professionals.

- **Mako** robotic systems have been used for more than 300,000 hip and knee replacement surgeries since 2006.[11] Based on data from CT scans and patient models, and using cameras mounted on its arm, the Mako robotic system maps the procedure and aligns the implant.

- A robot called **Moxi** has been designed to help nurses by carrying out the roughly 30% of nursing tasks that don't require patient interaction (such as dropping samples off at the lab for analysis). This frees up nurses to focus on patient care rather than running errands. But, interestingly, the robot has proven an unexpected hit with patients, who have asked for selfies and even sent fan mail. The team ended up programming extra activities for Moxi so that the popular bot could travel around the hospital even more, flashing heart eyes at patients as it goes.[12]

Robots in the Home

Going beyond the Roomba robot vacuum cleaner already mentioned, robots are beginning to find additional uses in our homes.

- The **LG Rolling Bot** is essentially a camera that can roll around your house and take pictures and videos, making it useful for keeping an eye on pets or monitoring security when you're away from home.

- **Zenbo** is like an all-in-one friend, babysitter, and remote control. The mobile companion robot can control household devices, share emotions, and keep the kids entertained by reading to them.

- **Dolphin** is a robotic pool boy that vacuums and scrubs your pool, intelligently deciding what functions are needed as it cleans.

Cobots in the Workplace

Many companies have been able to increase efficiencies and lower manufacturing costs with cobots.

- At the **Ford** Fiesta plant in Cologne, Germany, factory workers and cobots work alongside each other on assembly lines.[13]

- Remember I mentioned earlier in the chapter how, in **Amazon** fulfillment centers, cobots bring shelves of goods to human workers for packing? This has cut the time it takes to complete an order from one hour to just 15 minutes.[14]

- Online supermarket **Ocado** has a similar system in place. The human workers stay in one place while the cobots roam around picking groceries.[15]

Humanoid Robots

In the past, two-legged robots have proven to be a challenge for robotics companies because they're just not as stable as other robot designs. But that is changing.

- Boston Dynamics is at the forefront of robot agility. The company's **Atlas** robot is an impressive bipedal robot that can jump onto boxes, run, perform the odd backflip, and even indulge in a little parkour.[16]

- **Lynx** by UBTECH is designed to bring Amazon Alexa to life. Syncing with your Alexa, this humanoid robot can give you personalized greetings, play your favorite song, and give you the weather report.[17]

- The **Sophia** robot is so humanlike it's been granted citizenship by Saudi Arabia.[18] Apparently designed to look like Audrey Hepburn (but bald), Sophia has a sense of humor, can express feelings, and can communicate in an impressively fluid, intelligent way.

Robots Building Other Robots

Now, we even have robots capable of building other robots and repairing themselves.

- Swiss robotics company **ABB** is investing $150 million to build an advanced robotics factory in China that will use robots to build robots.[19]

- And thanks to 3D printing (see Trend 24), one Norwegian robot has learned to **self-evolve and 3D print itself**.[20]

The Weird and the Wonderful

Let's finish with some weird and wonderful examples that I couldn't resist including.

- The **RoboBee** X-Wing is a tiny robotic, solar-powered bee that could, in the future, help pollinate plants.[21]

- A 400-year-old temple in Japan has a **robotic priest**.[22]

- A French night club has unveiled **robot pole dancers** with CCTV cameras for heads.[23] It's like an art installation making a cheeky wink at robotics.

- We even have **robo-cockroaches** that are impossible to crush. Developed by a team at the University of California, Berkeley, the hope is the super-strong, tiny robot can be used in disaster relief.[24]

Key Challenges

Humankind created robots, and yet humankind has something of an innate fear of robots. Sure, most of us like the idea of a robot hoovering our dirty floors so we don't have to. But living or working with a humanlike robot would give many people pause for thought. In short, we like the idea of having certain tasks automated by robots, but we don't quite trust robots.

Therefore, businesses that want to implement robots will have to work hard to build trust between the human and the robotic workforce. This will mean being transparent with people about what processes will be automated and what this means for their jobs. But it also means selling the benefits of robots – such as robots taking on dull, repetitive tasks and freeing up human workers to focus on tasks that require greater skill. As with any new technology, it's much easier to get people's buy-in when they understand how the technology makes their working lives easier, better, and safer.

There will also be regulatory challenges to overcome as regulators place greater scrutiny on robots – particularly when it comes to autonomous machines gathering and using data. Business leaders can expect new regulatory frameworks to come into force to govern the use of robots.

There may also be specific physical challenges to overcome before installing robots in your workplace. For example, a human can move around on an uneven floor without any great difficulty. People are flexible and excel at adapting to their environment. But most robots would struggle to navigate an uneven floor. What you don't want is a situation where human workers are having to constantly stop what they're doing to help out their robotic colleagues! In many cases, robots may need to be customized to ensure they're suitable for organizations' individual needs and environments.

Then there's the cost factor, although thankfully the cost of robots is coming down, lowering the barriers to entry for businesses. The rapid growth of robots-as-a-service (RaaS) will also help to make robotic solutions more affordable and achievable for a wider range of businesses. RaaS is similar to AI-as-a-service (see Trend 1) or the software-as-a-service model that most businesses are familiar with. In essence, it allows companies to lease robotic automation services via a subscription service instead of having to pay for equipment outright or worry about maintenance – great for small- and medium-sized businesses looking to benefit from robotics. It also gives organizations the chance to scale up and down easily. It's no wonder then that industries as diverse as warehousing, healthcare, and security are beginning to benefit from RaaS and experiment with robotic solutions. Companies offering or developing RaaS solutions include Amazon (AWS RoboMaker), Google (Google Cloud Robotics Platform), and

Honda (Honda RaaS). In fact, research predicts that there will be 1.3 million installations of RaaS by 2026.[25]

How to Prepare for This Trend

Robotics provides exciting opportunities to reduce costs, increase capacity, boost efficiencies, and reduce errors. In the future, I believe humans will no longer be employed to carry out those jobs that robots can do safer, faster, more accurately, and less expensively.

What this means for your specific business will depend on your industry and everyday business processes. But I would advise all business leaders to begin thinking about how they could combine the unique capabilities of human workers with the efficiencies of robots to get the best out of both.

For example, humans are still more dexterous and better at coming up with unique, creative ways of solving problems. They're also empathetic and have emotional intelligence. Introducing more robots in the workplace will enable humans to do more of what they excel at – but this will inevitably require careful change management, training, and the acquisition of new skills.

As with all the trends in this book, technology is only moving in one direction: forwards. Whether you see robots as a tremendous opportunity or the beginning of the end for the human race, one thing is for certain: change is coming to your workplace. Those businesses that will benefit most from the robotics trend are the ones that can look for opportunities to bring robot workers and human workers together to enhance business success.

Notes

1. The future of robotics: 10 predictions for 2017 and beyond: www.zdnet.com/article/the-future-of-robotics/

2. IFR forecast: 1.7 million new robots to transform the world's factories by 2020: https://ifr.org/news/ifr-forecast-1.7-million-new-robots-to-transform-the-worlds-factories-by-20/

3. 2019: The year Nvidia gets serious about robots: https://thenextweb.com/artificial-intelligence/2019/01/14/2019-the-year-nvidia-gets-serious-about-robots/

4. Meet the cobots: humans and robots together on the factory floor, *Financial Times*: www.ft.com/content/6d5d609e-02e2-11e6-af1d-c47326021344

5. Starship Technologies raises $40 million for autonomous delivery robots: https://venturebeat.com/2019/08/20/starship-technologies-raises-40-million-for-autonomous-delivery-robots/

6. Nuro's Pizza Robot Will Bring You a Domino's Pie, *Wired*: www.wired.com/story/nuro-dominos-pizza-delivery-self-driving-robot-houston/

7. The rise of robots-as-a-service: https://venturebeat.com/2019/06/30/the-rise-of-robots-as-a-service/

8. Robot police officer goes on duty in Dubai: www.bbc.co.uk/news/technology-40026940

9. A robot cop that executes traffic stops. But will cops test it?: www.zdnet.com/article/a-robot-cop-that-executes-traffic-stops-but-will-cops-test-it/

10. Meet the Robot Firefighter That Battled the Notre Dame Blaze, *Popular Mechanics*: www.popularmechanics.com/technology/robots/a27183452/robot-firefighter-notre-dame-colossus/

11. Early Focus on Surgical Robotics Gives Stryker a Leg Up, *Forbes*: www.forbes.com/sites/jonmarkman/2019/08/30/early-focus-on-surgical-robotics-gives-stryker-a-leg-up/#1c542f822948

12. A hospital introduced a robot to help nurses. They didn't expect it to be so popular, *Fast Company*: www.fastcompany.com/90372204/a-hospital-introduced-a-robot-to-help-nurses-they-didnt-expect-it-to-be-so-popular

13. Ford tests collaborative robots in German Ford Fiesta plant: www.zdnet.com/article/ford-tests-collaborative-robots-in-german-ford-fiesta-plant/

14. Meet your new cobot: Is a machine coming for your job, *The Guardian*: www.theguardian.com/money/2017/nov/25/cobot-machine-coming-job-robots-amazon-ocado

15. Experimenting with robots for grocery picking and packing: www.ocadotechnology.com/blog/2019/1/14/experimenting-with-robots-for-grocery-picking-and-packing

16. Atlas: www.bostondynamics.com/atlas

17. Lynx robot with Amazon Alexa: www.youtube.com/watch?v=ocvWU bbx3GU

18. Saudi Arabia grants citizenship to a robot for the first time ever, *Independent*: www.independent.co.uk/life-style/gadgets-and-tech/news/saudi-arabia-robot-sophia-citizenship-android-riyadh-citizen-passport-future-a8021601.html

19. Robots will build robots in $150 million Chinese factory: www.engadget.com/2018/10/27/abb-robotics-factory-china/

20. Norwegian robot learns to self-evolve and 3D print itself in the lab, *Fanatical Futurist*: www.fanaticalfuturist.com/2017/01/norwegian-robot-learns-to-self-evolve-and-3d-print-itself-in-the-lab/

21. What Could Possibly Be Cooler Than RoboBee? RoboBee X-Wing, *Wired*: www.wired.com/story/robobee-x-wing/

22. This temple in Japan has a robotic priest: www.youtube.com/watch?v=4lTUDv4TX70

23. French Nightclub to Debut Robot Pole Dancers: https://interesting engineering.com/french-night-club-to-debut-robot-pole-dancers

24. Has science gone too far? This invincible robo-cockroach is impossible to squish: www.digitaltrends.com/cool-tech/cockroach-robot-withstand-massive-weight/

25. Manufacturing: How Robotics as a Service extends to whole factories: https://internetofbusiness.com/how-robotics-as-a-service-is-extending-to-whole-factories-analysis/

TREND 14
AUTONOMOUS VEHICLES

The One-Sentence Definition

An autonomous vehicle – be it a car, truck, ship, or other vehicle – is one that can sense what's going on around it and operate without human involvement.

What Are Autonomous Vehicles?

To explain how the technology works, I'll mostly focus on autonomous cars (also often described as *self-driving cars*). However, as you'll see later in the chapter, vehicles of all shapes and sizes are becoming increasingly autonomous.

Every major car manufacturer is investing heavily in self-driving technology of one sort or another. And while we're still a way off the fully autonomous cars seen in sci-fi films (where humans simply sit in the back and take it easy), we're getting surprisingly close to that becoming reality.

But what do we mean by "autonomous"? Autonomy in vehicles is classified according to the following levels:[1]

- Level 1: Basic driver support features that can provide steering *or* brake/acceleration support, such as lane centering or adaptive cruise control. So, essentially, only one driver process at a time is automated.

- Level 2: Still focused on driver support as opposed to automation, cars at this level can provide steering *and* brake/acceleration support at the same time. A self-parking feature would fall under level 2 autonomy.

- Level 3: Now we move into automated drive features that can drive the vehicle under limited conditions. A traffic jam chauffeur feature is a good example of level 3 autonomy. At this level, the car is capable of driving itself but, crucially, the driver must be ready to take the wheel when the system requests.

- Level 4: At this level, the vehicle will not require a human to take over, and it may not even be fitted with pedals or a steering wheel. However, a level 4 autonomous vehicle can still only operate under certain conditions – one example being a local driverless taxi that can only operate in a limited zone.

- Level 5: This offers the same level of autonomy as level 4, in that a human is not needed to take over. However, what separates level 5 from level 4 is that the autonomous features can drive the vehicle *anywhere and under all conditions*.

At the time of writing, we don't yet have commercially available level 4 or level 5 cars on the road, and most cars only offer level 2 autonomy. Even Tesla's highly autonomous cars – Tesla being one of the most advanced commercial auto manufacturers in this area – are considered level 2. But the race is on to develop truly autonomous cars for commercial sale. For example, Volvo has said it's aiming to have level 4 autonomous vehicles on the road by 2021.[2]

So how do self-driving cars work? It requires a lot of advanced technology to enable autonomous cars to understand what's going on around them, and decide what to do in response. Largely, this is possible thanks to sensors (see the IoT, Trend 2) and computer vision technology (Trend 12), but AI (Trend 1) and big data (Trend 4) also play a critical role. When it comes to sensors, radar is used to detect objects, including their size and speed. Lidar (light imaging detection and ranging) – which is similar to radar, but uses laser light pulses instead of radio waves – may also be used to map the car's surroundings. However, both radar and lidar are limited in that they don't really "see" what's going on around the car. This is where cameras come in. Self-driving cars are fitted with cameras that can read road signs, recognize road markings and provide an accurate view of the car's surroundings. Together, these technologies (along with others, like GPS) help the vehicle scan and map its surroundings, navigate a route, carry out maneuvers, and avoid obstacles.

There are many advantages to self-driving cars taking to the road, with the main benefit being improved road safety. Research has shown that driver error is by far the biggest cause of road traffic accidents,[3] due to factors like miscalculations, errors of judgment, speeding, drink-driving, and phone use. And when applied to public transport, autonomous vehicle technology could help authorities run more efficient transport networks. It's also hoped that autonomous vehicles will improve parking in congested cities, because driverless vehicles will simply drop passengers off and move on.

All things considered, self-driving vehicles could change the face of our cities. Congestion and pollution will decrease (in the future, it's expected that the majority of self-driving vehicles will be electric or hybrids). And land that is currently used for huge car parks could be repurposed for housing or public spaces. In fact, in Arizona, the city of Chandler has already changed its zoning laws to facilitate autonomous vehicles. Developers can build properties with fewer parking spaces so long as they provide suitable curb-side passenger loading zones.[4]

What's more, the autonomous vehicles of the future will drastically improve the daily commute, especially when we reach the point where human intervention is never needed. Instead of sitting behind the wheel, we'll be able to stretch out in the back, get ahead on some work, or simply relax. When you think that the average commute (in America, at least) amounts to 19 full working days a year, that's a lot of time commuters will be able to claw back for themselves.[5]

It's no surprise, then, that most traditional car manufacturers (plus some tech heavyweights) are chasing the self-driving car dream. More than 40 companies are actively investing heavily in autonomous technology and developing self-driving vehicles, from companies like BMW to Google parent company Alphabet.[6]

How Are Autonomous Vehicles Used in Practice?

Let's look at how different types of vehicles (not just cars) are becoming increasingly automated, and what we can expect of the self-driving vehicles of the future.

Self-Driving Cars

We might not be able to go to the showroom and buy a fully autonomous car today, but autonomous technology in cars is progressing rapidly.

- In 2018, **Volvo** unveiled a concept car that was part car, part hotel room, part office, and part flight cabin. The 360c represents Volvo's vision of a luxurious, driverless future, in which a car can pick you up and take you wherever you need to go. Interestingly, Volvo sees the 360c as a competitor for short-distance air travel.[7] Passengers could order a car (or "road plane" as Volvo sees it), pre-order food and drink, recline and relax in the back (maybe watch a movie), and arrive at their destination without the stress, exhaustion, and carbon footprint of plane travel. For journeys that could be made overnight, this vision could seriously disrupt the travel industry.

- Forget catching up on work or having a nap in the back of a self-driving car. **BMW** clearly has other benefits in mind. In 2019, the company released an ad portraying a bright future where people can have sex in self-driving BMWs. The ad was promptly deleted.[8]

- Chinese tech giant **Baidu** develops hardware and software for autonomous vehicles and claims its Apollo Lite vision-based tech solution – which uses 10 cameras to understand what's going on around the vehicle – delivers level 4 autonomy.[9] The list of companies Baidu is currently working with includes Ford, Volvo, and Hyundai.

- **Tesla** CEO Elon Musk famously promised that a Tesla autonomous vehicle would be able to drive coast to coast in the US without driver intervention by the end of 2017. That hasn't materialized yet, but Musk has said he expects Tesla cars to be able to operate without any driver intervention in 2020.[10] I believe that this seems ambitious, but we'll see if Tesla manages to pull it off.

Autonomous Taxis and Public Transport

In the future, fewer of us may actually own cars, preferring to hail a self-driving taxi or use autonomous public transport…

- **Waymo**, Alphabet's self-driving car company, has been trialing self-driving cars in Phoenix, Arizona, through its ride-hailing service Waymo One. As at the time of writing, all Waymo self-driving cars still have a human safety driver behind the wheel ready to take over if needed; however, in October 2019, Waymo sent an email to users of its ride-hailing app saying that "completely driverless Waymo cars are on the way."[11] The company has also been trialing autonomous taxis in California, where it transported more than 6,000 passengers in its first month.[12]

- In 2019, Chinese ride-hailing giant **Didi Chuxing** revealed it was launching its own self-driving pick-up service in Shanghai, and has plans to expand beyond China by 2021. The car will still have a human driver present.

- **Olli 2.0** is a 3D-printed (see Trend 24) autonomous shuttle vehicle made by Local Motors. With a top speed of 25 miles per hour, the vehicle is designed for campuses, military bases, hospitals, and other low-speed environments, rather than public roads. Olli 2.0 achieves level 4 autonomy.[13]

- A fleet of **Toyota** e-Pallette autonomous buses will transport athletes around the Olympic site at the 2020 Olympics in Tokyo.[14] The vehicles are five meters long, can carry 20 passengers at a time, and have a range of 100 miles. Toyota is also partnering with companies like Uber, Amazon, and Pizza Hut to explore how the vehicles could be used to deliver goods as well as people.

- Not to be outdone, **Volvo** has unveiled a 12-meter-long autonomous bus capable of carrying 93 passengers. The company says it is the world's first driverless electric bus.[15] Two of the buses are currently being trialed in Singapore: one at a university campus and the other at a bus depot.

Self-Driving Trucks and Vans

Autonomous trucks and vans promise to overhaul the transport industry.

- Autonomous vehicle startup **Gatik's** self-driving vans are being used to deliver Walmart groceries ordered online to local stores in Arkansas.[16] As yet, the vans still have a safety driver behind the wheel.

- **UPS** has partnered with self-driving truck startup TuSimple to transport cargo between Phoenix and Tucson in Arizona.[17] The trucks still have a safety driver and engineer on board.

- Currently, much of the focus in automated trucks is on "platooning," where vehicles travel closely together in a platoon, with a human driver in the lead vehicle. **Peloton Technology**, for example, has created an "automated following" system that means two trucks only need one driver.[18] The driver in the lead truck is in control and driving, while the truck following 55 feet behind is driverless.

- **Daimler Trucks** is partnering with tech firm Torc Robotics to test autonomous lorries on public highways in Virginia – with a safety driver and engineer on board, of course.[19]

Self-Driving Bikes and Scooters

Because, why have four wheels or more when you can have just two or three?

- The **REV-1** autonomous robot operates more like a bicycle than a car. With three wheels and a top speed of 15 miles per hour, the autonomous delivery vehicle is designed to operate in bike and car lanes. It's already being used to deliver food from two restaurants in Ann Arbor, Michigan.[20] (You might also like to read about delivery robots in Chapter 13.)

- **Segway-Ninebot** has unveiled a self-driving electric scooter that can drive itself to the charging station. The e-scooter is reportedly expected to take to the roads in 2020.[21]

- Now, we even have a bike that can drive itself, thanks to a special AI chip created by a team at **Tsinghua University** in China.[22]

Autonomous Ships

Just as on the road, the vast majority of maritime accidents are related to human error – as many as 75–96% of all maritime accidents, according to Allianz.[23] So it makes sense that our waterways will increasingly be sailed by autonomous vessels.

- The world's first fully autonomous car ferry, the result of a collaboration between **Rolls-Royce** and Finnish ferry company **Finferries**, was unveiled in 2018. Thanks to AI, the ferry can navigate and operate without any intervention from a human crew, although a land-based captain is able to monitor the voyage and, if necessary, take charge via remote control.[24]

- The world's first autonomous zero-emission container vessel, the **Yara Birkeland**, is already under construction and due for launch as early as 2020.[25]

- While some companies are busy building new autonomous ships, others are building technology that can be retrofitted into existing ships to make them more autonomous. One example is San Francisco-based startup **Shone**, which offers technology that can detect and predict the movement of other vessels on the water.

If autonomous vehicles on land and sea aren't enough for you, turn to Trend 19 to read about autonomous drones and aerial vehicles.

Key Challenges

There are a number of challenges to iron out before fully autonomous vehicles take to the road. Firstly, there are major legal obstacles. At the time of writing, the legal framework just isn't in place to govern autonomous vehicles on the road, leaving many critical things open to interpretation. Who is responsible if a fully autonomous, driverless vehicle is involved in an accident, for instance? To be able to evaluate this, we'll need a regulatory framework that sets out what is reasonable decision-making in autonomous vehicles; that way, if an autonomous vehicle is involved in an incident, investigators will be able to assess whether the vehicle acted within the set boundaries of good decision-making or whether the system failed. There are also many unanswered questions around how to insure autonomous vehicles.

Secondly, there are technological challenges to overcome, too, before we can reach full level 5 autonomy. Even the most advanced driver assistance features on the market today can incorrectly interpret their environment every now and then. Building the systems of the future that can interpret every possible scenario on the roads better and safer than a human driver is a significant challenge. Thirdly, there are also concerns around security – specifically, the concern that vehicles could be hacked. Overcoming these tech-related challenges in a car that's reasonably affordable for the everyday consumer? That's an even bigger challenge for car manufacturers! Keeping in mind the costs and economies of scale, it's likely that taxi/ride-sharing offerings, transport companies, and public transport providers will be the first to embrace fully autonomous vehicles, with everyday car owners adopting the technology (assuming it's affordable) further down the line.

Finally, as with many of the trends in this book, increasing automation will inevitably lead to job losses. Lorry drivers, taxi drivers, public transport drivers, and couriers are most at risk of losing their jobs to the autonomous vehicles of the future. Retraining and redeploying these workers will be vital.

How to Prepare for This Trend

The extent to which your organization will be impacted by this trend will depend on what type of business you run. Sectors like transport, logistics, and insurance may be significantly disrupted compared to other sectors.

In general, we can expect businesses to adopt autonomous vehicles earlier and on a wider scale than consumers, so it makes sense to start thinking about how your organization (and your competitors) might begin to incorporate autonomous vehicles into business processes. Whatever sector you operate in, this may include:

- Working with logistics providers who use driverless trucks.

- Using autonomous vehicles or delivery robots (see Trend 13) for customer deliveries.

- Using autonomous vehicles in depots and fulfillment centers.

- Employees interacting with customers' automated vehicles – for example, customers may send their automated vehicle to pick up that week's groceries.

You may also need to think about your company's physical use of space. In the future, if more of your employees and customers are traveling via driverless taxi or ride-sharing services, do you need to allow so much space for parking? Parking spaces will become far less important than drop-off and pick-up zones.

But, realistically, for most businesses, the shift to autonomous vehicles will be a gradual one. Rather than waiting years and upgrading an entire fleet of vehicles to fully autonomous ones, it's probably better to upgrade and adopt new autonomous features gradually. In other words, look for ways currently available technology can help you improve safety, deliver a better service to customers, and save costs *now*, as well as in the future.

Notes

1. SAE Levels of Driving Automation: www.sae.org/news/2019/01/sae-updates-j3016-automated-driving-graphic
2. By 2021, you could be sleeping behind the wheel of an autonomous Volvo XC90: www.digitaltrends.com/cars/volvo-xc-90-level-4-autonomy/

3. Traffic Safety Facts: https://crashstats.nhtsa.dot.gov/Api/Public/View Publication/812115

4. City planners eye self-driving vehicles to correct mistakes of the 20th-century auto, *The Washington Post*: www.washingtonpost.com/transportation/2019/07/20/city-planners-eye-self-driving-vehicles-correct-mistakes-th-century-auto/

5. Americans spend 19 full work days a year stuck in traffic on their commute, *New York Post*: https://nypost.com/2019/04/19/americans-spend-19-full-work-days-a-year-stuck-in-traffic-on-their-commute/

6. 40+ Corporations Working On Autonomous Vehicles: www.cbinsights.com/research/autonomous-driverless-vehicles-corporations-list/

7. Volvo's futuristic 360c concept is at once a hot-desk, hotel room and flight cabin: www.wallpaper.com/lifestyle/volvo-360c-autonomous-concept-car-review

8. BMW posts, deletes ad about sex inside self-driving cars: https://futurism.com/the-byte/bmw-ad-sex-self-driving-cars

9. Baidu claims its Apollo Lite vision-based vehicle framework achieves level 4 autonomy: https://venturebeat.com/2019/06/19/baidu-claims-its-apollo-lite-vision-based-vehicle-framework-achieves-level-4-autonomy/

10. Tesla's Musk Is Over-Promising Again on Self-Driving Cars, *Forbes*: www.forbes.com/sites/chuckjones/2019/10/22/teslas-musk-is-overpromising-again-on-self-driving-cars/#7bf081965e98

11. Waymo to customers: "Completely driverless Waymo cars are on the way": https://techcrunch.com/2019/10/09/waymo-to-customers-completely-driverless-waymo-cars-are-on-the-way/

12. Waymo's robotaxi pilot surpassed 6,200 riders in its first month in California: https://techcrunch.com/2019/09/16/waymos-robotaxi-pilot-surpassed-6200-riders-in-its-first-month-in-california/

13. Meet Olli 2.0, a 3D-printed autonomous shuttle: https://techcrunch.com/2019/08/31/come-along-take-a-ride/

14. This autonomous Toyota bus will carry athletes during the 2020 Tokyo Olympics: www.pocket-lint.com/cars/news/toyota/149705-this-autonomous-toyota-minibus-is-going-to-be-used-during-the-2020-tokyo-olympics

15. Volvo unveils "world's first" autonomous electric bus in Singapore: www.dezeen.com/2019/03/06/volvo-autonomous-electric-bus-design-singapore/

16. Gatik's self-driving vans have started shuttling groceries for Walmart: https://techcrunch.com/2019/07/27/gatiks-self-driving-vans-have-started-shuttling-groceries-for-walmart/

17. UPS has been quietly delivering cargo using self-driving cars: www.theverge.com/2019/8/15/20805994/ups-self-driving-trucks-autonomous-delivery-tusimple

18. This company created "automated following" so two trucks only need one driver: https://mashable.com/article/automated-following-peloton-autonomous-vehicles-trucking/?europe=true

19. Self-driving trucks are being tested on public roads in Virginia: www.cnbc.com/2019/09/10/self-driving-trucks-are-being-tested-on-public-roads-in-virginia.html

20. A new autonomous delivery vehicle is designed to operate like a bicycle, *The Washington Post*: www.washingtonpost.com/technology/2019/07/25/new-autonomous-delivery-vehicle-is-designed-operate-like-bicycle/?

21. Segway-Ninebot introduces an e-scooter that can drive itself to a charging station: www.theverge.com/platform/amp/2019/8/16/20809002/segway-ninebot-electric-scooter-self-driving-uber-lyft-charging-station

22. This autonomous bicycle shows China's rising expertise in AI chips, *Technology Review*: www.technologyreview.com/f/614042/this-autonomous-bicycle-shows-chinas-rising-ai-chip-expertise/

23. Shipping safety – human error comes in many forms: www.agcs.allianz.com/news-and-insights/expert-risk-articles/human-error-shipping-safety.html

24. Rolls-Royce and Finferries demonstrate world's first fully autonomous ferry: www.rolls-royce.com/media/press-releases/2018/03-12-2018-rr-and-finferries-demonstrate-worlds-first-fully-autonomous-ferry.aspx

25. Yara Birkeland Press Kit: www.yara.com/news-and-media/press-kits/yara-birkeland-press-kit/

TREND 15
5G AND FASTER, SMARTER NETWORKS

The One-Sentence Definition

5G is the fifth generation of cellular network technology, which together with other network innovations will give us much faster and more stable wireless networking, as well as the ability to connect more and more devices and enable richer, more varied streams of data.

What Are 5G and Faster Smarter Networks?

Networking technology may not seem as immediately fashionable or sexy compared to other innovations such as artificial intelligence, robotics, and automated vehicles, but it's the backbone of our online society and a smarter world. As bandwidth and coverage have increased, more has become possible, from email to web browsing, location-based services, and streaming video and games. Today it's all about sending a constant stream of real-time data back and forth between ourselves, the apps, and devices we use, and the cloud services which power them.

Faster data hasn't ever meant simply more data – it also means an increasing variety of data that opens up exciting possibilities for innovation. But as the number of cameras, scanners, and sensors collecting data from IoT devices (circle back to Trend 2 for more) increases, networks need to become smarter, as well as faster, to handle the new paradigms they will service.

The big driver this year is the arrival of the first consumer 5G networks. As well as greatly increased speed, the 5G networking protocols can cope with connecting far more devices within a geographical area – around 1 million devices packed into a square kilometer, rather than the 100,000 or so that 4G networks can handle.[1]

Advanced Services

Just as the availability of 4G and fiber-optic broadband connections ushered in the era of Netflix and streaming high-definition video, 5G has emerged as "cloud gaming" becomes a reality. Services such as Google's Stadia read the input from a player's game controller, relay it to servers in the cloud, and show the results on the player's screen at home, as seamlessly as if they were playing a game on a console placed under their television.

The extra speed of 5G will even make cloud VR (virtual reality, see Trend 8) a reality. Although modern VR headsets have become less bulky and more wieldy in recent years, they are still hampered by the need to either be wired to a computer or contain all the components needed to generate graphics internally.

With cloud VR over a mobile network, headsets will only need to function as screens – meaning they can become far lighter and more portable – while all of the processing needed to interpret the players' movements into actions in the virtual world is done remotely.

This promises to be revolutionary far beyond the realms of entertainment, as VR and augmented reality become increasingly adopted in industry, education, and healthcare. The power of advanced networking to make breakthrough technologies more widely available and easier to access, rather than simply faster, is key to the importance of this trend.

Cutting the Cord

5G is set to be the first widely available mobile networking technology that offers higher speeds than contemporary, easily-accessible consumer and business wired networking. This on its own means big changes, as infrastructure planners can do away with the need for tangled cables and inflexible access points in many situations. While switching to a cellular service might not be ideal for every operation – and the fastest landline speeds will surpass readily available 5G speeds for some time yet – it's likely to become a popular choice for data-driven organizations that need speedy and flexible networks.

Another useful feature of 5G networks is that they can be "sliced." This means that service providers can split their service into any number of virtual networks, each different in terms of functionality and performance. While running on 5G, systems as divergent as those needed to run industrial machinery, stream video to homes, and pilot driverless cars can take place on the same network, despite their vastly different requirements.[2]

5G networks also allow much more sophisticated "beamforming" than previous technologies. Beamforming is the process by which network transmitters and receivers direct their signals to the devices they communicate with.[3] This means that these networks will act far more reliably when it comes to communicating with fast-moving vehicles or passengers on public transport.

Smarter Networking

Just as speeds are increasing, networks are getting smarter too, thanks to new paradigms such as "mesh networking." This essentially means networks where every node is connected to every other node, and they can communicate directly. In traditional networks, large numbers of devices may all be connected through a single router or network adapter, making it a single point of failure as well as a speed bottleneck.

Satellite technology is undergoing a revolution as well – with the emergence of relatively affordable and accessible low-Earth orbit technology, meaning information can be more quickly and reliably spread to the farthest corners of the planet. In fact, one analyst recently told me that the cost to the average business of launching a satellite was quickly gaining parity with the average cost of launching a smart phone app – around $100,000.

And, of course, 5G won't be the end of the story. There will be 6G technology superseding even the capabilities of 5G networks. US president Donald Trump tweeted, "I want 5G, and even 6G, technology in the United States as soon as possible."

While he was mocked due to the fact that no one even knows what 6G will look like at this point, his wish is likely to be granted eventually. The group of researchers which defined the 5G standard met in Finland this year to start considering its successor,[4] although it is thought to be about 10 years away.

How Are 5G and Faster Smarter Networks Used in Practice?

In late 2018, the big service providers started rolling out the first consumer 5G mobile networks, although coverage is currently limited to major cities.[5] This is likely to change quickly as cheaper 5G handsets become available and more of us are able to get on board – in fact it's been forecast that most countries in the world will be able to access 5G services by the end of 2020.[6] Speeds are forecast to reach 1 gigabit per second – meaning a full HD movie can be downloaded in a matter of moments.

Streaming Entertainment

Google, **Microsoft**, **Sony**, and **Nvidia** have all either launched, or are in the process of launching, the next step forward for streaming entertainment – videogames – and these services will likely become more viable as network speeds increase. While streaming a movie simply involves a linear transfer of data, gaming requires images to be rendered and sent in response to input from the user, potentially also taking into account the action of hundreds of other online players. As network speeds increase, interactive experiences will become much more sophisticated and immersive.[7]

It isn't just the speed which will improve the user experience – the ability to handle many more simultaneous connections will hopefully put an end to the frustration of being unable to access mobile data in busy locations like city centers, supermarkets, and train stations.

Enabling an Autonomous World

High-speed mobile data networks have clear implications for operating autonomous vehicles and robots. **Toyota** has demonstrated the first 5G-enabled humanoid robot, T-HR3. Before the deployment of 5G, the robot required a wired data connection, as mobile data transfer speeds were not fast enough for situations where the robot has to be controlled remotely by humans.[8]

Today, 5G networks are an integral part of plans to roll out self-driving cars. Roadside sensors as well as the vehicles themselves will communicate over high-speed mobile networks, managing the vehicle-to-vehicle as well as vehicle-to-cloud communications needed for safe navigation. It has been reported that switching to 5G for these communications, rather than 4G, will potentially reduce network latency from 20 milliseconds to 1 millisecond – which could be the difference between life and death when it comes to avoiding a collision.[9]

Meanwhile, in Germany, **Thales** and **Vodaphone** are carrying out the first trials of driverless trains connected by 5G networks, with the ability to "slice" the networks cited as essential. Using a 4G network there would never be total confidence that a control connection to the train could be maintained, due to the limit on the number of devices which can be connected within an area.[10]

Facilitating Better Healthcare

In China, which still has a huge part of its population living in relatively isolated areas, 5G is being used to expand coverage to people who have so far missed out on the online revolution.

One of the problems it is helping to tackle is the lack of doctors and medical professionals in remote areas. 5G networks are playing a part in telemedicine, allowing doctors to examine patients and, soon, carry out surgery from hundreds of miles away.

Assistant head doctor Zhou Yang, of the **Third People's Hospital of Chengdu**, said, "There is no lag time at all in the discussion between doctors. We've been wanting to do this for a while, but we could only do it now because of 5G technology."[11]

Smarter Networks

Another key mobile networking technology, known as Narrowband Internet of Things (NB-IoT) has been put to use by Chinese farmers for monitoring and tracking the location and health of their yak herds. **NB-IoT** has been developed to enable devices which can operate in very remote conditions for lengthy periods of time without needing maintenance. The information on the wandering herds allows remote diagnosis of illness or injury, as well as more efficient management of land to prevent overgrazing.[12]

Another interesting animal-centric use can be seen in the network of devices used in Korea to deter wild boars from roaming close to sites used during the recent **Olympic Winter Games**, by playing recordings of tiger roars.[13]

Amazon has thrown its considerable muscle behind the concept of mesh networking by integrating it into its upcoming **Sidewalk** consumer IoT system. Sidewalk will use low-bandwidth communications to build networks between connected devices within an area, which become more robust as more and more devices are connected. It aims to provide a middle ground between low power networks such as home wi-fi and Bluetooth, and high-powered cellular networks. The first apps to be linked in this way will be a range of smart pet tracking collars to be launched in 2020.[14]

Also due to launch in 2020 is a fleet of 200 miniature satellites weighing just 10 kg each, which will be used by British telco **Sky** and **Space Global** to carry voice and text data around the globe, including up to three billion people living in equatorial regions of Africa and South America who are currently poorly served.[15]

Meanwhile, Elon Musk's **SpaceX** is building a network of over 4,000 satellites that will form a network called **StarLink**, which it is claimed will offer coverage to the entire world.[16]

Key Challenges

Many of the challenges around integrating impending 5G networks will concern teething problems – it's a very new technology that in some cases has been rushed to market by multinational telcos in order to keep up with, or beat, their competition.

This is clear from the way it is currently only available in major cities, but also due to some underlying fundamentals to how current services operate. Current 5G networks often still rely on switching to data streams carried by older Long-Term Evolution networks to carry out functions such as authentication.[17]

This means that network speeds will inevitably be stalled by bottlenecks – although this problem is likely to disappear as more networks are upgraded to full 5G capability.

For many organizations, a decision that will need to be taken in the near future is whether to rely on 5G or existing wi-fi networks for high speed, local networking. Early experiences of 5G have shown that the signals are more prone to disruption from physical infrastructure such as tall buildings, or simply being inside, than existing technology. How you balance this with the greatly increased speed will mean carefully considering the specific needs of a system.

Initiatives like OpenRoaming[18] are likely to mitigate against the need to make these choices – allowing devices to seamlessly switch between different available networks in order to maintain the best possible performance.

Accessing 5G networks is currently also an expensive proposition, with carriers charging a premium for connection, and compatibility somewhat limited to higher-end consumer devices such as the very latest mobile phone handsets. Again, this will almost certainly change as adoption increases.

For business, the challenge is to work out how to benefit not just from the higher speed, but from the enhanced connectivity and sophisticated features like network slicing that 5G enables. As with all new technology, if your only strategy is to use it to do what you currently do, but faster, you are likely to lose out to more innovative competitors using it to build entirely new processes and business models.

And, of course, as with all new technology, it wouldn't be sensible to ignore the security threats it could potentially expose us to. The power and speed of 5G networks mean that sophisticated security measures such as encryption, anonymization, and virtualization can be easily integrated into streams as standard. However, the security of any network is only as strong as its weakest point, and the proliferation of connected devices it will enable means hackers looking for a potential access point will have more options to choose from.

How to Prepare for This Trend

My key piece of advice here would be not to simply think about how 5G and other advanced networking technology can increase your speed, but to consider the new possibilities it might bring.

As always, it's important to start by considering an overall strategy – rather than simply understanding everything that 5G and other advanced networking technology can do, and shoehorning it in. What do you need 5G and advanced networking to do for you in order to meet your current aims?

A recent Accenture survey of business leaders found that over half did not expect that 5G would allow them to do anything new or understand how they could use it to build new services.[19] As we've seen, following the introduction of previous new networking technology such as 3G and wi-fi, this is likely to be a mistake.

The jump from streaming movies and music to streaming interactive, immersive video games is a good example of this. Faster networking means less friction on the channels you use to communicate with your customers. Imagine a chatbot that can instantly open a live video stream and demonstrate how to solve a problem that a customer is having using your product or service.

Most organizations hoping to benefit will also have to take the opportunity to overhaul their IT systems. Local infrastructure will need the capability to quickly scale its ability to communicate with wireless or remote devices in order to reap the full benefits, and security implications should be addressed at the earliest opportunity.

It's also important to start making sure that the possibilities these new technologies bring are understood at every level of an organization. Giving employees enhanced abilities to collaborate and communicate will bring its own rewards and board-level buy-in is essential for getting the go-ahead on big projects.

New networking technologies will also greatly boost connectivity to more parts of the world, opening up access to new markets, as well as new talent. Identifying ways to capitalize on this will be a key challenge in coming years.

Notes

1. 1 Million IoT Devices Per Square Km – Are We Ready for the 5G Transformation?: https://medium.com/clx-forum/1-million-iot-devices-per-square-km-are-we-ready-for-the-5g-transformation-5d2ba416a984
2. What Is Network Slicing?: https://5g.co.uk/guides/what-is-network-slicing/
3. What is beamforming, beam steering and beam switching with massive MIMO: www.metaswitch.com/knowledge-center/reference/what-is-beamforming-beam-steering-and-beam-switching-with-massive-mimo
4. Why 6G research is starting before we have 5G: https://venturebeat.com/2019/03/21/6g-research-starting-before-5g/
5. 5G Has Arrived in the UK And It's Fast: www.theverge.com/2019/5/30/18645665/5g-ee-uk-london-hands-on-test-impressions-speed
6. 5G Availability Around the World: www.lifewire.com/5g-availability-world-4156244

7. Cloud Gaming: Google Stadia and Microsoft xCloud Explained: www.theverge.com/2019/6/19/18683382/what-is-cloud-gaming-google-stadia-microsoft-xcloud-faq-explainer
8. DOCOMO and Toyota Conduct Successful Remote Control of T-HR3 Humanoid Robot Using 5G: www.nttdocomo.co.jp/english/info/media_center/pr/2018/1129_01.html
9. Why 5G Is The Key to Self-Driving Cars: www.carmagazine.co.uk/car-news/tech/5g/
10. Thales and Vodafone conduct driverless trial using 5G: www.railjournal.com/signalling/thales-and-vodafone-conduct-driverless-trial-using-5g/
11. How China is Using 5G to Close the Digital Divide: https://govinsider.asia/connected-gov/how-china-is-using-5g-to-close-the-digital-divide/
12. Mobile IoT Connects China to the Future: www.gsma.com/iot/news/mobile-iot-connects-china-to-the-future/
13. Who's winning the global race to offer superfast 5G?: www.bbc.co.uk/news/business-44968514
14. Amazon Sidewalk is a new long-range wireless network for your stuff: https://techcrunch.com/2019/09/25/amazon-sidewalk-is-a-new-long-range-wireless-network-for-your-stuff/
15. The Low Cost Mini Satellites Bringing Mobile to the World: www.bbc.co.uk/news/business-43090226
16. SpaceX is in communication with all but three of 60 Starlink satellites one month after launch: www.theverge.com/2019/6/28/19154142/spacex-starlink-60-satellites-communication-internet-constellation
17. T-Mobile relies on LTE for 5G launch: www.lightreading.com/mobile/5g/t-mobile-relies-on-lte-for-5g-launch/a/d-id/754355
18. OpenRoaming explained: https://newsroom.cisco.com/feature-content?type=webcontent&articleId=1982135
19. Business and Technology Executives Underestimate the Disruptive Prospects of 5G Technology, Accenture Study Finds: https://newsroom.accenture.com/news/business-and-technology-executives-underestimate-the-disruptive-prospects-of-5g-technology-accenture-study-finds.htm

TREND 16
GENOMICS AND GENE EDITING

The One-Sentence Definition

Genomics is an interdisciplinary field of biology that focuses on the understanding and manipulation of DNA and genomes of living organisms. Gene editing is a group of technologies that enables genetic engineering in order to change the DNA and genetic structure of living organisms.

What Is Genomics and Gene Editing?

Our understanding of the human genome has continued to increase since it was first accurately sequenced in 2003. In large part this is due to the increasingly powerful computers and sophisticated software tools that have become available.

A Quick Biology Refresher

All living cells contain DNA that determines the traits that a cell will pass on when it divides. This DNA can be thought of as a "code" which governs the production of the new proteins that are built as living cells divide. As we have a better understanding of how sequences of DNA (genes) are passed on during that division process, we can have a better idea of the impact it will have on a living organism's ability to cope with injury, allergies, food intolerances, hereditary diseases, or any number of internal or external factors. The study of this field of science is known as genomics.

Manipulating Our DNA

Taking that a step further, biotechnology is advancing to the point where it's viable to alter the DNA encoded within a cell. This will influence the characteristics or traits (phenotypes) that its descendants will have after it reproduces by cell division. This alteration can be done, for example, by removing – by physically cutting – sections of DNA from a strand. The strand will then naturally heal, and pass on an "altered" version of the DNA when it divides, meaning the new cell will develop with altered characteristics.

In plants, this could affect the number of leaves or their coloring, while in humans it could affect their height, eye color, or their likelihood of developing diabetes. Gene editing, in plants or animals, is particularly useful if "bad" genes are detected that could endanger the health of the organism or its descendants.

This opens up a range of possibilities that are almost unlimited, as it means that any characteristic of a living organism that is inherited can theoretically be changed. Children could be made immune to illnesses that their parents are susceptible to, crops can be developed that are resistant to pests and diseases, and medicines could be tailored to individuals according to their own genetic makeup.

CRISPR Gene Editing

One method of gene or genome editing in particular, known as CRISPR-Cas9,[1] first developed in 2012 at the University of California

Berkeley, has been shown to have tremendous potential. This method has been described as practically an "off-the-shelf" solution for gene editing,[2] making advanced bioscience like gene editing a possibility outside of academia and big business.

The microscopic scale that targeted gene editing needs to function at is truly amazing, given that the human body contains around 37 trillion cells. The nucleus, where most DNA resides, takes up around 10% of the mass of a typical cell. The level of accuracy needed to cut something so small is inconceivable to most people, but with CRISPR-Cas9 it is accomplished by programming a "guide" RNA to match with the specific place where a DNA strand needs to be cut, and using it to deliver an enzyme (Cas9) that will split the strand.

In humans, gene editing is generally carried out outside of the body for safety reasons – cells are extracted before the DNA is cut, and then reinserted.

The development of these new methods means that there is currently a great deal of competition among academic researchers and private enterprise to develop new gene-based solutions to problems facing us and the world – from improving our day-to-day health to creating new agriculture and livestock science to feed the hungry. The value of the global market in genomics is forecast to rise from $14 billion to $32 billion between 2017 and 2025, as these technologies, services, and products become more widely available.[3]

How Is Genomics and Gene Editing Used in Practice?

Much of the work being done with gene editing is in the field of healthcare. Here, its potential applications are almost unlimited, but among the most exciting current projects is "correction" of DNA mutations which can lead to serious illnesses such as cancer or heart disease.

- This means that babies born at a high risk of developing these conditions can have their ability to withstand damage caused by mutation and their resistance to hereditary illness increased. One example is work being done to eradicate a mutation known as **MYBPC3**, which is recognized to cause a heart disease called hypertrophic cardiomyopathy that affects as many as one in 500 adults and is a leading cause of sudden death.[4]

- Gene therapy work carried out to prevent disease or improve health can be classified into two types – germline therapy, which can cause changes in reproductive cells (eggs and sperm) and therefore cause changes that will be inherited by offspring, and somatic, which targets non-reproductive cells and can potentially cure or slow down the spread of disease in the target organism.[5]

- Other diseases where progress has been made include **Duchenne muscular dystrophy** – a devastating condition that affects one in 3,500 young boys and results in early death. Here, gene editing has been demonstrated to fix the mutation when it occurs in beagle dogs,[6] leading to hope that a treatment for humans will not be far behind.

- Other animal-centric research is aimed at improving the health of pets and companion animals themselves, for example by eradicating hereditary traits such as a predisposition towards blindness, bladder stones, and heart defects within particular breeds of dogs.

- Elsewhere, gene editing has led to the development of "**super horses**" – modified to run faster and jump higher. The first results of this work, developed by Argentinian company **Kheiron Biotech**, are due to be born in 2019 and could reduce the need for highly expensive breeding programs traditionally used in racehorse breeding.

- Even without the need for gene editing, genomic data can be beneficial to **public health** and in particular the development of **precision medicine**, where treatment is based on an individual's genetic makeup. In Ireland, genetic data on patients with epilepsy is being included in electronic health records, which has led to an increased understanding of the genetic causes behind the condition.[7]

- Plant health can be improved with gene editing, too. Vegetable and cereal crops such as wheat, soybeans, and rice are all susceptible to pests and disease, often necessitating the use of chemical fertilizers and pesticides which have knock-on effects on the environment. By editing plant genomes, their resistance to these threats can be increased, leading to increased yields and less dependence on harmful chemical interventions. With the world heading for a global food crisis brought about by growing populations and climate change,[8] this technology could prove vital for feeding future generations.

- Chocoholics will be excited to hear of one application of gene editing which aims to solve several problems that severely limit the amount of chocolate that can be created worldwide. Researchers at Penn State University are working on creating **genetically enhanced cacao** trees which will be resistant to the disease and fungus that destroys 30% of the worldwide cacao crop before their pods can be harvested.[9] This is done by suppressing a gene which decreases the plant's ability to fight off infections. As well as increasing the global supply this could also hugely improve the livelihoods and living conditions of cacao farmers, who are some of the most deprived agricultural workers.

- As well as increasing yield, crops can also be developed which are tastier and more visually appealing. This is expected to have a knock-on effect on human health as people become increasingly likely to choose fruit or vegetable-based snacks over unhealthy fast and convenience food. Foods can also be developed that respond better to freezing, increasing their shelf-life and reducing food loss caused by natural decay.

- UK startup **Tropic Biosciences** have demonstrated a gene-edited method for creating a naturally caffeine-free coffee bean, which it says could drastically reduce the cost of decaffeinated coffee, as well as increase the nutritional benefit to humans. It is also working on creating disease-resistant bananas – tackling the problem of banana growers currently dedicating one quarter of their production expenses towards fungicides and pesticides.[10] With both bananas and coffee beans among the most widely consumed foodstuffs across the globe, their work could have far-reaching consequences.

- Another possibility is eliminating the dangers to humans caused by allergens. Compounds and substances within foodstuffs such as cereals, dairy products, and nuts responsible for allergic reactions can potentially be eliminated through gene editing. In one project, researchers at **Wageningen University**, Netherlands, are removing antigens in gluten from wheat, making it healthy for those with a gluten intolerance[11] – who are often restricted to eating gluten-free products which can be more expensive and of lesser nutritional benefit.

Key Challenges

Perhaps more so than with any other trend in this book, there are a huge number of ethical and legal concerns as well as "what if?" questions when it comes to genetic manipulation and editing!

As gene editing brings the possibility that man made changes to the genome will be passed down through future generations – potentially

impacting the future evolution of many species, including humans –
work is highly regulated.

Germline editing in humans – which produces results that will be
carried down through generations by tampering with reproductive
cells – is currently banned in many countries, including much of
Europe, as its long-term results are not understood. This will possibly
change in coming years as public discussion on the ethics and implications advances, or the need for eradicating disease becomes more
urgent.

In late 2018, a scientist based at the Chinese Southern University of
Science and Technology caused controversy with claims that he had
used gene editing to cause twin girls to be born with an inbuilt immunity to the HIV virus. This was done legally, as germline editing is not
banned in China (or, as it happens, in the United States). However,
the international scientific community was quick to respond that the
process is still far from being considered fully safe and there could be
unforeseen consequences. Another objection could be that there was
no real benefit, as there are other "safe and effective" ways to protect
against HIV.[12]

When it comes to genetic modification of food crops (genetically modified (GM) foods), there is also an understandably high
level of public concern. The majority of European Union nations[13]
(including the United Kingdom), as well as Russia, ban the general cultivation of GM crops, although they allow the importation
of GM foods from other countries – particularly for use as animal feed. Other nations, including China, Japan, and Canada, allow
GM foods to be cultivated but they are subject to strict rules and
regulations.

Other countries where GM foods are permitted to be grown (with
varying degrees of regulation) include the United States, Brazil, Aus-
tralia, India, and Spain.

An important distinction is that crops created through gene editing techniques such as CRISPR-Cas9 are not considered to be genetically modified organisms (GMOs) in the US, as they are not developed by mixing genes from different organisms. In theory, the changes that are made could come about through natural evolutionary processes, such as natural selection. In Europe, regulators do not agree – and gene-edited crops were classified as GMOs by a court decision in 2018.

It's also true that, as with other major advances in healthcare, much of the impact of genomics and gene editing is limited to richer countries in the developed world. Ensuring that the arrival of this technology doesn't lead to a widening of the gap in standards of healthcare between developed and developing nations, and that as many people will benefit as possible, is a key challenge.

In addition, as our understanding of genomics increases, there is a danger that personal information in the wrong hands could have negative consequences. It's hard to think of data that is more personal than an individual's genetic makeup, and there's a danger that it could be used, for example, to deny health insurance to people or groups considered to be at risk. If "designer babies" – highly resistant to illness and programmed for a long, healthy life – become available to the rich, but not the poor, society could become segregated along the lines of genetics, furthering inequality.

This brings up the issue of expense – gene editing has traditionally been an expensive (as well as unpredictable) process. However, the introduction of CRISPR-Cas9 has gone a long way towards reducing the costs to a level that is affordable for many organizations.

How to Prepare for This Trend

On a personal level, it's already possible to take advantage of the increase in genomics knowledge, and new genomics technology, by gaining a better understanding of our own bodies and our own specific genetic makeup. An increasing number of private companies, such as 23andMe, offer personalized genetic testing that can provide details ranging from your ancestry to your disposition to health conditions, including diabetes and Alzheimer's and Parkinson's diseases. They can also highlight whether you are likely to pass on genes to your descendants that will make them predisposed to hereditary conditions such as cystic fibrosis and sickle cell anemia.

For businesses and organizations looking to genomics and gene editing to provide new business models, processes, or simply to increase their competitive edge, then an understanding of the regulatory environment they are operating within is essential. As well as this, knowledge of the public attitudes, which vary across cultures and geographical areas, is key to understanding how your operations may be affected, as this trend becomes more widely implemented.

As with anything as potentially transformative to society as genomics, it can be easy to get carried away thinking about possibilities such as wiping out cancer, or even indefinitely prolonging human life. In reality, such huge advances are still likely to be a long way away, if they are ever even possible at all. Focusing on solving simpler problems that will have an immediate real-world impact is likely to be more fruitful in the short term. Often, as with building immunity to HIV, there are solutions that are achievable without resorting to expensive and largely unknown genomic technological solutions, and these avenues should be thoroughly explored first.

Notes

1. What is CRISPR-Cas9?: www.yourgenome.org/facts/what-is-crispr-cas9
2. Gene Editing, an Ethical Review: http://nuffieldbioethics.org/wp-content/uploads/Genome-editing-an-ethical-review.pdf
3. Genomics Market Growing Rapidly With Latest Trends & Technological Advancement by 2027: https://marketmirror24.com/2019/07/

genomics-market-growing-rapidly-with-latest-trends-technological-advancement-by-2027/

4. Correction of a pathogenic gene mutation in human embryos: www.nature.com/articles/nature23305
5. How is Genome Editing Used?: www.genome.gov/about-genomics/policy-issues/Genome-Editing/How-genome-editing-is-used
6. Gene editing restores dystrophin expression in a canine model of Duchenne muscular dystrophy: https://science.sciencemag.org/content/362/6410/86
7. Integrating Genomics Data Into Electronic Patient Records: www.technologynetworks.com/genomics/news/integrating-genomics-data-in-to-electronic-patient-records-322634
8. Climate Change and Land: www.ipcc.ch/report/srccl/
9. Cocoa CRISPR: Gene editing shows promise for improving the "chocolate tree": https://news.psu.edu/story/521154/2018/05/09/research/cocoa-crispr-gene-editing-shows-promise-improving-chocolate-tree
10. This startup wants to save the banana by editing its genes: www.fastcompany.com/40584260/this-startup-wants-to-save-the-banana-by-editing-its-genes
11. CRISPR Gene Editing Could Make Gluten Safe for Celiacs: www.labiotech.eu/food/crispr-wageningen-gluten-celiac/
12. Genome-edited baby claim provokes international outcry: https://www.nature.com/articles/d41586-018-07545-0
13. Several European countries move to rule out GMOs: https://ec.europa.eu/environment/europeangreencapital/countriesruleoutgmos/

TREND 17
MACHINE CO-CREATIVITY AND AUGMENTED DESIGN

The One-Sentence Definition

Machine co-creativity and augmented design refers to the ability of machines to be creative, specifically to enhance the work of humans in creative and design processes.

What Is Machine Co-Creativity and Augmented Design?

As we've already seen in this book, artificial intelligence (AI) (Trend 1) is now giving machines the ability to replicate a range of human functions, including:

- Reading and writing (natural language processing and generation, Trend 10)

- Understanding speech and conversing (voice interfaces and chatbots, Trend 11)

- Seeing (machine vision and facial recognition, Trend 12)

There's no doubt that machines are now capable of being incredibly intelligent and productive. But there's one trait we've always seen as uniquely human: producing art and other creative endeavors. This ability to imagine and bring to life something that previously didn't exist can't possibly be matched by machines. Or can it?

In fact, machines are already taking on more and more creative tasks that we wouldn't have thought possible even just a few years ago. The lines between what humans can make and what machines can make are becoming increasingly blurred. For example, as machines have got better at understanding text (natural language processing), their ability to *create* new text (natural language generation) was the logical next step. Thanks to AI, machines can now create all sorts of text, from news articles and company reports to entire books. It's the same with images. Back in 2012, Google had trained its AI to recognize cats in YouTube videos. Since then, the ability of machines to interpret images has developed rapidly, and now, having learned to understand images, machines can create new images that didn't exist before (see examples later in the chapter).

One of the first really impressive examples of machine creativity came in 2016 when Google's AlphaGo AI program beat world champion Lee Sedol at the ancient Chinese board game Go. This may not seem that impressive when you think that computers have been beating humans at chess since 1996. But the early chess-playing computers weren't strictly intelligent or creative – they just used pure computing brute force to consider every possible chess move in existence. But the AlphaGo system did something different; it came up with an entirely new Go move that had never been seen before.[1] Considering Go is famed as being a game that relies on creativity and intuition, AlphaGo's feat was seriously impressive.

In his book *The Creativity Code: Art and Innovation in the Age of AI*, Marcus du Sautoy argues that art is surprisingly mathematical, made up of lots of patterns and structures.[2] Yet, very often those patterns are hidden. AI is great at unearthing hidden patterns, learning from those patterns, and applying those patterns in new ways – and this is what enables machines to be "creative." But even du Sautoy admits that the creative heavy lifting tends to be done by the person programming the system, not the system itself.

This brings us to the major sticking point in machine creativity. As yet, machines struggle to replicate true human creativity because there's much we don't understand about the human brain's creative thought processes. Those inspired ideas that seemingly come out of nowhere? Those "aha!" moments that stop us in our tracks? We've yet to understand in any meaningful way how that magical and mysterious process works. Therefore, typically, machines in the creative process have to be "told" by humans what to create before they can produce the desired end result. In other words, for now at least, machine creativity is being largely used to augment and enhance the human creative process. This is what we mean by *co-creativity* or *augmented design* – deploying AI alongside human creativity, rather than replacing human creativity altogether. Think of it as a little extra creative "muscle," if you will.

This extra creative muscle is useful for humans because, just like machine creativity, human creativity has its limitations. As American chemist Linus Pauling – the only person to have won two unshared Nobel Prizes – put it, "You can't have good ideas unless you have lots of ideas." But, while humans may excel at making sophisticated decisions and pulling ideas out of thin air, we're not great at producing a vast number of possible options and ideas to choose from. In fact, we tend to get overwhelmed and less decisive the more options we're faced with! This is where co-creativity pays dividends. Machines have no problem coming up with infinite possible solutions and permutations, and then narrowing the field down to the most suitable options – the ones that best fit the human creative's "vision." In this way, by combining machine creativity with human creativity, it's possible to create completely new things that may never have been possible for humans or machines to create alone.

Generative design is one example of the combined power of human and machine creativity. This cutting-edge field involves enhancing the work of human designers and engineers using intelligent software. Very simply, the human designer inputs their design goals, specifications, and other requirements, and the software takes over to explore all possible designs that meet that criteria. Generative design could be utterly transformative for many industries, including architecture, construction, engineering, manufacturing, and consumer product design. I touch on some specific examples of generative design later in the chapter.

How Is Machine Co-Creativity and Augmented Design Used in Practice?

Let's look at some of the ways in which machines are able to enhance the creative process.

Visual Arts

- In 2016, **IBM's Watson** AI platform was used to create the first ever AI-generated movie trailer. As the trailer was for horror movie *Morgan*, the system analyzed hundreds of existing horror movie trailers, looking at visuals, sound, and composition. Based on what it learned, Watson selected suitable scenes from Morgan for editors to compile into a trailer, thereby condensing a weeks-long process into just one day.[3]

- **Christie's** made the headlines in 2018 when it became the first auction house to sell a work of art created by an AI algorithm. The painting, called *Portrait of Edmond de Belamy* sold for an incredible $432,500 – nearly 45 times its estimate.[4]

- An AI has been developed that can bring old paintings "to life," creating realistic animated versions of famous artworks like the

Mona Lisa. In a widely circulated video, the Mona Lisa can be seen looking around and moving her lips.[5]

- An AI called StyleGAN has been used to create images of people that do not exist. The faces may look real, but they're entirely fake. Check out the faces for yourself at **thispersondoesnotexist.com**. The faces are considered to be some of the most realistic-looking AI-generated images of non-existent people so far.

- If you fancy collaborating with AI on your own work of art, check out the **Deep Dream Generator** tool.[6] It can transform any image you upload and create a new one following a particular art style. You can even get your collaborative artwork printed. Or, for something a little weirder, the **Dreamscope** app takes your perfectly normal images and warps them into nightmarish, trippy new images.[7]

- One AI can even create images of food, just based on reading the recipe. Developed by computer scientists at **Tel Aviv University**, the system was trained with 52,000 recipes and real-life images of food, after which it learned to generate synthetic images from new recipes.[8] The resulting pictures are pretty mixed (some of them don't look like things I'd like to eat), but in terms of the technology, it's pretty cool.

Music

- Musician and composer David Cope developed a system called **EMI**, or **Experiments in Musical Intelligence**, to help him compose music and overcome "composer's block." It works by uploading existing compositions, which are then analyzed to identify patterns that signify the piece's style. Based on its analysis, EMI could then recombine the various elements into new patterns without duplicating anything existing. Thanks to EMI,

Cope says he began to recognize patterns in his own compositions that he wasn't aware of, and this encouraged him to change up his style.[9]

- **AIVA** is an AI that can compose emotional soundtrack music, and is designed to help creatives put music to their projects in a faster, easier way. You can compose with preset styles (including pop, rock, and, erm, sea shanty), or create something based on your influences.[10]

- Tech companies are developing tools to help artists create music with AI. For example, **Google's Magenta** project has produced songs written and performed by AI.[11]

- A grammy-nominee producer has used IBM's Watson AI platform to help create new music. **Alex da Kid** used Watson to analyze the composition of five years' worth of hit songs, as well as cultural reference points like newspaper articles, film scripts, and social media commentary, in order to understand the "emotional temperature" of the time period and suggest certain "themes" for a song. After that, da Kid used the Watson BEAT platform to come up with musical elements that fitted the chosen theme.[12]

- There's even a virtual popstar called **Yona** – with an AI-created voice, sound, and social media presence. The majority of Yona's lyrics, melodies, voice, and chords are computer-generated, although a human producer mixes and produces the final song.[13]

Dance

- Award-winning choreographer **Wayne McGregor** has used AI to suggest new choreography. The project – a collaboration with Google Arts & Culture Lab – demonstrates how AI can be used to help human creatives break known habits and patterns and suggest multiple new options that fit within their particular style. The algorithm was trained on thousands of hours of McGregor's videos, spanning 25 years of his career, and it used this data to predict 400,000 McGregor-like sequences.[14] McGregor said the tool "gives you all of these new possibilities you couldn't have imagined."

Generative Design

While examples of AI-generated music, dance, and art may be eye-opening, they may not be of much use to the average business leader. Generative design, however, could have a transformative impact on any company that designs and builds products, equipment, machinery, buildings, and so on. Earlier in the chapter, I mentioned how generative design or augmented design means using AI to make multiple design suggestions and enhance the work of human designers. The beauty of generative design is that the software does all the heavy lifting of working out what works and what doesn't – giving companies much greater choice of designs, and saving them the time and expense of creating prototypes that don't deliver.

Let's look at some examples of this in practice:

- In a collaboration between software company Autodesk and renowned designer **Philippe Starck**, generative design was used to create a new chair design. Starck and his team set out the overarching vision for the chair and fed the AI system questions like "Do you know how we can rest our bodies using the least amount of material?" From there, the software came up with multiple suitable designs to choose from. The final design – an award-winning chair named "A.I." – debuted at Milan Design Week in 2019.[15]

- **NASA** has used generative design to come up with a concept for a spider-like interplanetary lander.[16] Lighter and slimmer than most previous NASA landers, the new design could be used to explore distant moons like Europa.

- **General Motors** has used generative design software to redesign a seatbelt bracket – replacing a fiddly eight-part assembly with one single part that's 40% lighter and 20% stronger.[17]

- **Airbus** deployed generative design to come up with thousands of variations for cabin partitions – ending up with a design that was half the weight of the previous design, and saving millions of dollars in fuel costs in the process.[18]

The Written Word

Here are just a few examples of how AI is already creating written content, but if you circle back to Trend 10 there are many more examples of natural language generation in action.

- AI has been used to create an almost award-winning novel called *The Day a Computer Writes a Novel*. The novel made it through the first round of screening in a Japanese national literary prize.[19] A team from the **Future University Hakodate** in Japan set certain parameters, words, and sentences for the AI, before letting the system "write" the novel itself.

- **Kogan Page** has published a book written with the help of AI. Called *Superhuman Innovation*, it's the first book about AI to be written by a human author and AI author working together.[20]

Key Challenges

For now, as I've mentioned, replicating true human creativity is a major challenge. Until we understand fully how the human creative thought process works, AI is unlikely to achieve true creativity. However, as the examples in this chapter show, AI is already proving a valuable support tool for creatives.

As I see it, the biggest challenge around machine co-creativity is finding the right balance between humans and machines, and figuring out how best to capitalize on the strengths of both. Where humans excel is usually in coming up with a creative vision, connecting with their target audience, and making complex decisions on which design (or song, or artwork, or whatever) is most likely to resonate with that audience. AI can support this process by coming up with multiple options that fit the determined style or parameters – quicker, easier, and more effectively than a human could. Interestingly, this collaborative process can inspire people to go in new directions that they might never have considered before.

Organizations that use AI to enhance their creative processes will no doubt have to overcome a certain amount of fear and skepticism from the human workforce – and perhaps from clients and end users, as well. Overcoming this is a matter of communicating the benefits of co-creativity, so that your human team can become advocates for AI-enhanced creativity.

How to Prepare for This Trend

For now, the applications of this trend in business are relatively limited. However, particularly for businesses that incorporate aspects of design, AI could provide a significant boost to the design process.

If you're interested in exploring machine co-creativity and augmented design in more detail, you might like to read *The Creativity Code: Art and Innovation in the Age of AI* by Marcus du Sautoy, which is a really interesting exploration of AI creativity.

When considering potential applications, remember that it's about using AI to complement and enhance the work of humans – not to replace human creativity altogether. It's about finding ways for humans and AI to work together to come up with something more incredible than either could create alone.

Notes

1. How AI is radically changing our definition of human creativity, *Wired*: www.wired.co.uk/article/artificial-intelligence-creativity?utm_medium=applenews&utm_source=applenews
2. *The Creativity Code: Art and Innovation in the Age of AI* by Marcus du Sautoy, 2019, Harvard University Press
3. IBM Research Takes Watson to Hollywood with the First "Cognitive Movie Trailer": www.ibm.com/blogs/think/2016/08/cognitive-movie-trailer/
4. Is artificial intelligence set to become art's next medium?: www.christies.com/features/A-collaboration-between-two-artists-one-human-one-a-machine-9332-1.aspx
5. "Mona Lisa" Comes to Life in Computer-Generated "Living Portrait". *Smithsonian*: www.smithsonianmag.com/smart-news/mona-lisa-comes-life-computer-generated-living-portrait-180972296/
6. Deep Dream Generator: https://deepdreamgenerator.com/
7. Create your own DeepDream nightmares in seconds, *Wired*: www.wired.co.uk/article/google-deepdream-dreamscope
8. AI created images of food just by reading the recipes, *New Scientist*: www.newscientist.com/article/2190259-ai-created-images-of-food-just-by-reading-the-recipes/
9. EMI: When AIs Become Creative And Compose Music: https://bernardmarr.com/default.asp?contentID=1833
10. AIVA: www.aiva.ai/
11. Google Magenta: https://magenta.tensorflow.org/
12. Grammy Nominee Alex Da Kid Creates Hit Record Using Machine Learning, *Forbes*: www.forbes.com/sites/bernardmarr/2017/01/30/grammy-nominee-alex-da-kid-creates-hit-record-using-machine-learning/#4e0010062cf9
13. Speaking to Yona, the AI singer-songwriter making haunting love songs: www.dazeddigital.com/music/article/40412/1/yona-artificial-intelligence-singer-ash-koosha-interview
14. Could Google Be The World's Next Great Choreographer?, *Dance Magazine*: www.dancemagazine.com/is-google-the-worlds-next-great-choreographer-2625652667.html
15. From Analog Ideas to Digital Dreams, Philippe Starck Designs the Future With AI: www.autodesk.com/redshift/philippe-starck-designs/
16. AI software helped NASA dream up this spider-like interplanetary lander: www.theverge.com/2018/11/13/18091448/nasa-ai-autodesk-jpl-lander-europa-enceladus-artificial-intelligence-generative-design

17. Think Generative Design is Overhyped? These Examples Could Change Your Mind: www.autodesk.com/redshift/generative-design-examples/
18. Think Generative Design is Overhyped? These Examples Could Change Your Mind: www.autodesk.com/redshift/generative-design-examples/
19. A Japanese A.I. program just wrote a short novel, and it almost won a literary prize: www.digitaltrends.com/cool-tech/japanese-ai-writes-novel-passes-first-round-nationanl-literary-prize/
20. Kogan Page publishes book about AI, written with the help of AI: www.koganpage.com/page/kogan-page-publishes-book-about-ai-written-with-the-help-of-ai

TREND 18
DIGITAL PLATFORMS

The One-Sentence Definition

A digital platform is a mechanism or network that facilitates valuable connections and exchanges between people – and these exchanges may include communicating and sharing information, selling products, or offering services.

What Are Digital Platforms?

Facebook, Uber, Amazon, and Airbnb are all well-known examples of digital platform businesses. What do they all have in common? They aid valuable interactions between people, meaning participants may use the platform to sell goods or services to each other, collaborate on projects, give advice, share information, or cultivate friendships.

Platform businesses have existed for years. If you think about it, a shopping center is a platform to link consumers to those who make and sell clothing, shoes, and other goods. Likewise, a newspaper is a platform that connects advertisers to readers. Platforms aren't new. What is new about today's most powerful platform businesses is that the connections they facilitate take place in the online, rather than the physical, world, and are enabled by data. In other words, related trends like mobile devices, artificial intelligence (Trend 1), big data (Trend 4), cloud computing (Trend 7), and automation (Trend 13) have combined to create a perfect storm, giving rise to a new wave of highly successful digital platform businesses. These platforms are also fueling big changes in how we live and work, giving rise to the gig economy and sharing economy.

Via a digital platform, users get easy, on-the-go access to the people, products, and services that interest them. And in return for making those connections happen so easily, the platform gets access to a wealth of data on its users' preferences and habits – data that helps the platform business improve its service offering and keep the community coming back for more.

The digital platform approach has turned the traditional business model on its head. Where a traditional business might place much of its value in its physical assets and raw materials, the value of a digital platform lies not in what it owns internally, but how well it can leverage an external ecosystem. As an example, compare a global hospitality company like Marriott with Airbnb. Marriott's business model is based heavily on assets, buying or building hotels that then need to be run and maintained by huge teams of employees. Airbnb, however, taps into the unlimited power of the crowd to connect travelers with those who have a place to stay. Airbnb leverages a global travel ecosystem without having to build and run a single hotel.

For Airbnb, the platform *is* the business. Success doesn't lie in how nice its hotels are or how slick the hotel service is; success lies in adding value for the platform user – by saving them money, eliminating traditional industry gatekeepers like big hotel chains and travel agents, providing a means for people to make money on their empty apartment, etc. This is true of most platform businesses; they're rarely providing actual goods or services to people. Instead, they act as a facilitator for the crowd, making interactions possible, easy, and safe for both the provider and the user.

As we've seen from companies like Uber and Airbnb, the more valuable a platform is to the people who use it, the more successful it becomes. In fact, seven of the 10 most valuable companies in the world are now platform businesses – and estimates suggest more than 30% of global economic activity could be mediated by digital platforms by 2025.[1] Across every industry, we're likely to see power shift away from companies with a traditional pipeline business model – whereby the company creates a product or service and funnels it to the intended market – towards companies that embrace the platform model – using platforms to leverage their industry's ecosystems and communities.

How Are Digital Platforms Used in Practice?

I've mentioned a few examples of digital platforms already, but let's delve a little deeper into real-life examples of platforms in action.

Social media networks like **Instagram** and **Twitter** are, at heart, digital platform businesses. **Google** is another example of a digital platform – after all, it connects people searching for stuff with advertisers who have stuff to sell. Online retail platforms like **Amazon** and **eBay** are also connecting consumers with the things they want to buy – as are booking platforms like **Booking.com**, **Skyscanner**, and **Expedia**. Then you have gig economy and sharing economy platforms like **Uber**, **Airbnb**, and **Upwork**.

If that sounds like platform businesses are purely a Silicon Valley invention, think again. Many powerful platform businesses have emerged outside of the US, particularly from China. For example:

- Alibaba Group absolutely dominates Chinese e-commerce, particularly through its **Taobao** platform, the world's biggest e-commerce website and, at the time of writing, the ninth most visited site in the world.[2]

- **Didi Chuxing** is China's leading ride-sharing app (in part due to its acquisition of Uber China in 2016),[3] and has even moved into the bike-sharing space.[4]

- Didi may be in for some competition though; online Chinese food delivery giant **Meituan-Dianping** launched its own ride-hailing service in 2018.[5]

So far, all of the examples I've given have centered around tech companies and innovative startups – businesses that have built themselves around the platform model from the start. But many established, non-platform businesses are also beginning to capitalize on the platform business model, either by creating digital platforms of their own or partnering with an existing platform provider to leverage an external ecosystem.

Let's look at some of the other ways more traditional companies are building digital platform business models to complement their existing offering:

- From the success of businesses like Uber and Didi, it's clear that "transportation as a service" presents a huge opportunity for car manufacturers. So it makes sense that **Volkswagen** is exploring ways to capitalize on this mobility-as-a-service movement.[6] Likewise, **Nissan** is in talks with Didi Chuxing in China to create a ride-sharing service centered on electric vehicles.[7] And in 2016, **Toyota** invested in car-lending app Getaround (like Airbnb, but for cars), and began integrating Getaround's technology into its vehicles, allowing Getaround users to unlock cars without a key.[8]

- Medical equipment manufacturer Philips is another example of an asset-based business that's adopting a platform model. The company has launched the **Philips HealthSuite** digital platform – a suite of tools that facilitate personalized healthcare.[9]

- Rail transit company Siemens Mobility, part of Siemens AG, has created the **Easy Spares Marketplace** – a platform that brings together manufacturers, dealers, and customers, allowing users to order all the spare parts they need in just one place.[10]

- General Electric has created the **Predix Platform**, designed to help GE's industrial clients collect and analyze data from their industrial equipment.[11]

- John Deere has created the **MyJohnDeere** platform, which helps farmers better manage, run, and maintain their agricultural equipment.[12]

Key Challenges

Creating a successful platform isn't easy. Many platform endeavors fail, and fail quickly. In fact, researchers looked at failed platforms and discovered that the average lifespan of a defunct platform is less than five years.[13] The project looked at 252 platforms and identified four common reasons why 209 of them failed:

- **Pricing incorrectly on one side of the market.** A platform often needs to subsidize one side of the market to encourage people to use it (for example, by pricing products low or charging minimal commissions) – think of how Amazon established itself quickly by aggressively discounting books. But which side of the market should be charged, and how much? Get that decision wrong and your platform may not survive for long.

- **Failing to develop trust with platform users.** Building trust through rating systems, secure payment systems, and governance policies is essential for platform success. Because if users don't trust a platform, they'll go elsewhere.

- **Dismissing the competition.** Just because you're first into a market or overtake another platform to become market leader doesn't mean you'll stay number one. Many platforms (indeed, many businesses) fail because they get complacent about their position. The researchers point to Microsoft Explorer – which once captured almost 95% of the browser market but has since been overtaken by Firefox and Chrome – as a prime example.

- **Entering the market too late.** Even if you have a great platform, if you enter the market years after your competition, you'll very likely struggle to get a foothold. Being early into a market presents its own challenges, but being late is worse.

In addition to these four issues, there's another challenge that I see on the horizon for platform businesses: the rise of blockchain technology (Trend 6) and its potential to disrupt the platform model. Let's take Uber as an example. Uber may be confident in its position as the leading ride-sharing platform, but what if people who need a ride could connect directly with drivers, without the need for an intermediary platform? That's a role that blockchain could potentially fulfill in the future.

Platforms like Uber act as an aggregator or a centralized meeting place that connects providers with the people who need their products or services. It may feel decentralized when you're booking a ride and see that David or Ali is on their way and will arrive in two minutes – it feels like a transaction between you and your driver. But Uber owns or controls all the means by which that transaction takes place, including the software, servers, payment system, operating conditions, and service agreements. In other words, when you hail a ride through Uber, you're paying Uber. Then Uber pays your driver (after taking their cut of the fare). Without Uber, there's no easy way for you to connect with that particular driver in that particular moment.

Blockchain has the potential to change that. Circle back to Chapter 6 for an explanation of how the technology works, and you'll see that blockchain functions as an extremely secure, decentralized system. There is no centralized authority to dictate conditions and take its cut of the fees. People who need a ride could transact directly with drivers via a safe, secure, and trustworthy peer-to-peer system.

Skeptical that it could work in reality? The Arcade City blockchain-based ride-sharing app – which was created in response to one driver's frustration with Uber's way of working – is already available. And when it first launched, 3,000 drivers in 30 cities rushed to sign up, causing the app to suspend driver recruiting.[14] So perhaps the question isn't whether blockchain will disrupt platform businesses like Uber, but *when*.

How to Prepare for This Trend

As the examples in this chapter show, platforms offer growth opportunities across all kinds of businesses, sectors, and industries, not just for tech companies. That's why I believe every company can and should have a platform strategy – from small businesses and nimble new startups to large corporations with a more traditional pipeline business model.

But because platforms can often represent a fundamental change to business models and strategy, we're not talking about an easy, overnight transition. Rather, you'll have to think carefully about how best your business might leverage the platform model to drive success.

I recommend starting with the following questions:

Where's the Value?

Creating or capturing value is an essential part of building a successful platform. Therefore, a good starting point is to ask how your company could create or add value, or support the exchange of value, through a platform or network. In other words, how would your intended users benefit, both from the platform itself and by connecting with others in the platform?

This doesn't necessarily mean abandoning your existing business model. Rather, there might be opportunities to create a platform-powered additional revenue stream. For example, if you manufacture farming equipment, you could build a marketplace platform for spare parts, or a network that facilitates servicing and repairs – connecting your customers with value-added services from other providers.

Do You Have the Platform Skills and Knowledge You Need?

Be honest here, because many traditional corporations simply don't have the expertise – or even the culture – to seamlessly adopt a platform business model. You may need to look outside the company to the world of entrepreneurs and tech startups, potentially creating a joint venture to leverage the skills you need.

How Will You Attract People to Your Platform?

Without people, platforms fail. How long do you think Uber would last if it didn't have an army of drivers ready and waiting to respond to user requests for a ride? Likewise, Facebook relies on its community of users to generate and post the content people want to read and see. Just as Airbnb relies on attracting people with homes and rooms to rent. The community is critical to the success of your platform. You therefore need to work out how you'll "seed" users to your platform – and this may involve offering free services, low prices, reduced commission, or a uniquely specialized offering.

How Will Your Platform Encourage and Support Interactions between Users?

Platforms are all about leveraging a community or ecosystem. Ultimately, the platform itself should become the core of that community – it needs to be *the* place where consumers or users connect with people who can provide the information, goods, or services they need. That means the platform needs to encourage and facilitate valuable interactions between participants. To preserve the value of those interactions and ensure users continue to have a great experience on the platform, you'll need to develop some governance policies that make it clear what is and isn't acceptable.

How Will Your Platform Integrate Future Technologies?

As we've already seen, blockchain technology may threaten the very foundation of some of the existing first wave of platform businesses, like Uber. For any business moving into platforms now, this provides an opportunity to leapfrog existing platforms and become leaders in the next wave of platforms – ones that harness and benefit from new technologies like blockchain. So, at the outset, think about how your platform may overcome or even integrate this technology to create or add value; for example, is there an opportunity to use blockchain to create a new, more decentralized way of doing business in your industry?

One final tip when considering a platform model: don't try to mimic what's already out there. Sure, many businesses have succeeded by positioning themselves as "Uber for [insert product here]," but I believe it's important to develop a platform strategy that highlights your unique value proposition as a business. To put it another way, what is it you do best, and how could a platform model help you build on that success?

Notes

1. The Platform Economy, *The Innovator*: https://innovator.news/the-platform-economy-3c09439b56
2. The top 500 sites on the web: www.alexa.com/topsites
3. Confirmed: Didi buys Uber China in a bid for profit, will keep Uber brand: https://techcrunch.com/2016/08/01/didi-chuxing-confirms-it-is-buying-ubers-business-in-china/

4. Didi Chuxing declares war on China's bike-sharing startups: https://techcrunch.com/2018/01/09/didi-declares-war-on-chinas-bike-sharing-startups/

5. China ride-hailing war seen erupting again with new challenger to Didi, *South China Morning Post*: www.scmp.com/tech/start-ups/article/2135282/chinas-meituan-takes-didi-ride-hailing-expansion-set-trigger-new

6. #2 Platform Business Model – Mobility As A Service: https://platformbusinessmodel.com/2-platform-business-news-mobility-service/

7. Didi Chuxing Proposes Joint Venture With Nissan & Dongfeng, Seeks Capital Injection From SoftBank: https://cleantechnica.com/2019/07/02/didi-chuxing-proposes-joint-venture-with-nissan-dongfeng-seeks-capital-injection-from-softbank/

8. Toyota partners with Getaround on car-sharing: https://techcrunch.com/2016/10/31/toyota-partners-with-getaround-on-car-sharing/

9. Philips HealthSuite digital platform: www.usa.philips.com/healthcare/innovation/about-health-suite

10. Easy Spares Marketplace: https://easysparesmarketplace.siemens.com/

11. GE Predix Platform: www.ge.com/digital/iiot-platform

12. MyJohnDeere: https://myjohndeere.deere.com/mjd/my/login?TARGET=https:%2F%2Fmyjohndeere.deere.com%2Fmjd%2Fmyauth%2Fdashboard

13. A Study of More Than 250 Platforms Reveals Why Most Fail, *Harvard Business Review*: https://hbr.org/2019/05/a-study-of-more-than-250-platforms-reveals-why-most-fail?utm_medium=email&utm_source=newsletter_weekly&utm_campaign=insider_not_activesubs&referral=03551

14. Arcade City Is a Blockchain-Based Ride-Sharing Uber Killer: www.inverse.com/article/13500-arcade-city-is-a-blockchain-based-ride-sharing-uber-killer

TREND 19
DRONES AND UNMANNED AERIAL VEHICLES

The One-Sentence Definition

Drones, also known as unmanned aerial vehicles (UAVs), are aircraft that are piloted either remotely or autonomously.

What Are Drones and Unmanned Aerial Vehicles?

In 2013, Amazon CEO Jeff Bezos made headlines when he predicted drones would be delivering orders within five years – an idea that had more than a few people scoffing at the time. The company later came up against some stiff regulatory hurdles, which meant the prediction didn't quite come true on time. But that vision of parcels flying through the air to customers' doors is now becoming incredibly close.

Drones take many forms, from the small, inexpensive drones flown by enthusiastic hobbyists, to the multimillion-dollar military drones that have become a critical part of missions. Drone technology is continually evolving and the cutting edge involves developing autonomous drones that are powered by artificial intelligence (AI) (see Trend 1); as we'll see in this chapter, advanced military drones are now being developed that can act autonomously, completing their mission without human intervention.

I'll get to specific examples of drone use later in the chapter, but it's safe to say drones have found a wide variety of applications beyond military and hobby use. Just for starters, they're frequently used in mapping, aerial surveying, and search and rescue operations.

A typical drone is equipped with technology like GPS sensors (read more about sensors in Trend 2, the Internet of Things), gyroscopes, accelerometers, infrared cameras, first-person view cameras, and lasers. Many smaller drones are technically *quadcopters*, meaning they have four rotors, can take off and land vertically, and hover – so, they're more like a helicopter than a plane. Heavy duty military drones, on the other hand, tend to resemble small planes, in that they have fixed wings and require a runway to take off and land.

Drones typically come with a "return to home" feature, whereby if the drone loses contact with its controller or is low on battery, it will automatically return to its home point. The latest drones also come with systems that can detect and avoid obstacles while in flight. Increasingly, many drones are also being fitted with a "no fly zone" feature to prevent them flying into restricted areas. This will help to avoid repeats of the high-profile case in December 2018, where multiple drone sightings led to London Gatwick Airport's runway being closed for more than 30 hours and the flights of around 140,000 passengers being disrupted. You can read more about some of the challenges around drones later in the chapter, but for now, let's focus on the practical applications.

How Are Drones and Unmanned Aerial Vehicles Used in Practice?

Drones have so many potential uses. I believe the technology will transform how goods are delivered, and, ultimately, perhaps even transform how humans travel. Let's take a look at some awe-inspiring (and occasionally disturbing) real-world examples of drones in action.

Military Drones

There are many ways drones are used in military operations, from gathering intelligence to deploying armed drones against suspected terrorists.

- One key area of development is AI-powered drone swarms. These are groups of self-organizing drones that are capable of acting in unison, making decisions among themselves, and communicating with each other to achieve a set goal. In 2018, the US Defense Advanced Research Projects Agency (**DARPA**) confirmed it had developed a squad of drones that was able to "adapt and respond to unexpected threats ... with minimal communication."[1] That basically means when communications with the human controller were disrupted, the drones were still able to work together to achieve the mission's objective, without live human intervention. The UK government has confirmed that British forces will use similar drone swarms in the future.[2]

- **DARPA** has also been experimenting with the combination of drone swarms and ground robots (see Trend 13) for military missions. A test conducted in 2019 demonstrated how, in the future, ground-based robots and drone swarms could accompany infantry units in urban environments, and help military personnel to find, surround, and secure buildings. This could potentially involve up to 250 drones and robots at a time.[3]

- Meanwhile, the **US Army** has been trialing the use of tiny, palm-sized drones, which can be used to fly on ahead of soldiers and send back information, including video.[4]

- A video released by the **Russian Defense Ministry** shows its latest stealth combat drone, the S-70 Okhotnik-B, in action.[5] The drone is shown taking off and flying next to a combat jet, demonstrating how UAVs can accompany pilots on missions and help them improve the view of what's going on below. (Some military drones can also be used as decoys, drawing fire away from the main fighter jet.)

Search and Rescue and Firefighting Drones

Drones can be very useful during natural disasters and search and rescue missions.

- Researchers at the **University of Zurich** have developed a foldable drone, designed specifically for use in disaster zones, that can alter its shape to get through cracks and small spaces.[6]

- Drones are frequently used to help firefighters assess risk and danger, find people trapped in buildings (thanks to thermal cameras), create maps of buildings for fire planning purposes, and conduct fire investigations. But drones can also help to actively fight fires. **Aerones's** firefighting drone can withstand extreme heat and reach great heights to fight fires.[7]

- Drones are also becoming common in search and rescue missions. They can be used to search areas, or simply light up deserted areas at night to aid rescuers. In one example from the UK, a drone was used to help **Lincolnshire Police** find a man who had been thrown from his car during an accident. The accident happened on a cold night, and police were worried the man could die from hypothermia, so they sent up a drone equipped with a thermal camera, which quickly located him.[8]

Drones for Law Enforcement

Drones can be used in a number of law enforcement scenarios, from gathering video evidence to chasing suspects to assessing situations remotely before sending officers in.

- In one case where a man was holed up in a hotel and threatening to set off a grenade, the police were able to use a drone to **identify the grenade as a fake**.[9] This information was vital for the on-the-ground sniper team, who had been considering using lethal force. When they knew the grenade wasn't real, they were, thankfully, able to taser the man.

- **Traffic police in China** have used drones to relay orders to drivers who are breaking traffic laws. In one amusing example, aired on state TV, traffic police used a drone to tell a moped rider to put his helmet on.[10] The motorist duly complied.

- Drones can also play a valuable role in the fight against poaching. For example, the **Sea Shepherd Conservation Society** uses drones to catch poachers at work on the open seas.[11]

Delivery Drones

Jeff Bezos's vision of drones delivering goods to customers is now much closer to becoming reality…

- In June 2019, **Amazon** confirmed it would be launching self-piloted drones within a matter of months.[12] The drones – which use computer vision (see Trend 12) and machine learning (see AI, Trend 1) to fly and avoid obstacles such as power lines – can fly up to 15 miles and carry goods weighing up to 2.3 kg.

- Also in 2019, **Wing Aviation** (a subsidiary of Google parent company Alphabet) secured approval from the Federal Aviation Administration (FAA) to start making commercial drone deliveries in Blacksburg, Virginia.[13] Wing's drones had made more than 70,000 test flights prior to approval.

- **UPS** has also won FAA approval to operate a fleet of delivery drones. Initially, the unmanned package delivery service will be used to deliver packages to hospital campuses, but the company has plans to expand beyond that.[14]

- In remote areas, delivery drones could make the difference between life and death. In parts of **Rwanda and Ghana**, drones have been used to deliver blood and vital medical supplies.[15]

Industry Examples of Drones

From monitoring agricultural land to carrying out building surveys, drones have found a wide range of uses across different industries.

- In farming, drones can now be used to round up cattle and assess crops. In one example, a **French farming cooperative** used drones to better assess and treat crops, resulting in a 10% average increase in yields.[16]

- In **construction**, drones are making structural inspections safer and easier than ever. They can be used to inspect roofs and exteriors in far more detail than traditional aerial photographs, inspect bridges that have been built over large bodies of water, and inspect high-rise structures with ease.[17]

- **Airbus** has launched an innovative maintenance drone that promises to cut aircraft inspection times and improve the quality of inspection reports.[18] The automated drone follows a set inspection path inside the aircraft and captures images, which are then transferred to a central system for analysis. An inspection report is then automatically generated.

Passenger Drones

You might be surprised to know that several companies are already working on passenger drones. Could this be the answer to traffic congestion problems in densely populated cities like Los Angeles? Time will tell.

- German aviation company **Volocopter** has already conducted several tests of its two-seater 18-rotor air taxi, which can be operated by a pilot or on its own and runs entirely on electricity.[19] It can, at the time of writing, fly for 30 minutes and has a maximum range of 17 miles.

- **Uber** has its own plans to develop and deploy a flying taxi service by 2023.[20] Likewise, Airbus is aiming to have its **CityAirbus** electric passenger drone offering fully up and running by 2023.[21]

- Chinese startup **Ehang** is aiming to be the first to start regular flights on autonomous, pilotless passenger drones. The company could be running three or four regular flight routes in the city of Guangzhou as early as 2020.[22]

Key Challenges

For me, there are a number of ethical questions around military use of autonomous drones. Should we even want to develop drone swarms that can make tactical decisions for themselves, for instance? In theory, this means they could identify targets and deploy weapons without human intervention – an idea that makes me and many others very uncomfortable. A number of leading AI and robotics experts have signed an open letter calling for a ban on autonomous weapons, including autonomous drones.[23] Perhaps that's partly why, in 2019, the Pentagon was seeking to recruit an ethicist to oversee military AI.[24]

There are also concerns around security, specifically the possibility that drones could be hacked into. Protecting drones from such attacks will become all the more important, particularly for military drones that are capable of deploying weapons. (EU Security Commissioner Julian King has also warned about the threat of terrorists using drones to attack crowded spaces and mass gatherings.[25])

And there are more than a few regulatory wrinkles to iron out. In the US and UK, there are regulations in place for the non-recreational use of drones, which govern the size of drones, and speed and height they can travel at. Certain companies, such as Amazon, have secured exemptions from these rules to allow them to test delivery drones. But if commercial use of drones is to become more widespread – for example, with multiple different companies flying potentially thousands of drones above our cities – these regulations will no doubt need to be expanded. We'll need a full framework to govern the safe operation of commercial drones. Noise pollution and privacy will be other factors to consider, since drones are basically noisy, flying computers with cameras. (Interestingly, data from NASA shows that people find noise from drones particularly annoying compared to ground traffic.[26])

Decisions will also need to be made regarding air space, and how all these drones will interact with other drones – whether passenger drones need to be able to communicate with delivery drones, for instance. Air management systems will need to be significantly overhauled to cope with the rise in air traffic.

Regulation for passenger drones will be an interesting area to keep an eye on. If companies like Amazon have found it difficult to get permission to fly delivery drones, just imagine how much more difficult it'll be for passenger drones. (For good reason, obviously.) It's one thing to start testing unmanned passenger drones, but it's quite another to consistently send them up with test passengers on board.

There are physical infrastructure challenges to overcome as well. All these drones will need places to land, take off, and hang out in between deliveries or passenger pick-ups. As a result, the commercial use of drones may end up changing how our cities look.

And finally, something that's been mentioned as a challenge in many other chapters in this book, there's the risk to jobs. If delivery drones become the norm for certain goods, that will significantly impact the work of drivers (see also autonomous vehicles, Trend 14).

How to Prepare for This Trend

Clearly, the extent to which this trend will affect your business depends on which industry you're in. Those in the transport and logistics industry will need to begin thinking about the impact drones will have on their core business processes sooner rather than later. But really, any company that needs to ship goods (and people) from A to B may find this process can be enhanced through drones in the future.

Notes

1. CODE Demonstrates Autonomy and Collaboration with Minimal Human Commands: www.darpa.mil/news-events/2018-11-19
2. How swarming drones will change warfare: www.bbc.com/news/technology-47555588
3. Watch DARPA test out a swarm of drones: www.theverge.com/2019/8/9/20799148/darpa-drones-robots-swarm-military-test
4. Watch DARPA test out a swarm of drones: www.theverge.com/2019/8/9/20799148/darpa-drones-robots-swarm-military-test
5. Watch Russia's combat drone fly next to a fighter jet: https://futurism.com/the-byte/russia-unmanned-combat-air-drone-jet
6. Self-folding drone could speed up search and rescue missions: www.cnbc.com/2019/02/18/self-folding-drone-could-speed-up-search-and-rescue-missions.html
7. Aerones firefighting drone: www.aerones.com/eng/firefighting_drone/
8. Drones in Search and Rescue: 5 Stories Showcasing Ways Search and Rescue Uses Drones to Save Lives: https://uavcoach.com/search-and-rescue-drones/
9. "Eyes in the Sky" and Embry-Riddle Training Help Police End Stand-off. Embry-Riddle Aeronautical University: https://news.erau.edu/headlines/eyes-in-the-sky-and-embry-riddle-training-help-police-end-hotel-standoff
10. Police drone caught barking orders at Chinese driver: https://futurism.com/the-byte/police-drone-orders-chinese-driver
11. We Really Can Stop Poaching. And It Starts With Drones, *Wired*: www.wired.com/2016/07/we-really-can-stop-poaching-and-it-starts-with-drones/

12. Amazon drone deliveries to begin "in months", *Independent:* www.independent.co.uk/life-style/gadgets-and-tech/news/amazon-drone-deliveries-where-when-date-a8946566.html
13. Drone delivery taking off from Alphabet's Wing Aviation: www.therobot report.com/drone-delivery-taking-off-from-alphabets-wing-aviation/
14. UPS wins first broad FAA approval for drone delivery: www.cnbc.com/2019/10/01/ups-wins-faa-approval-for-drone-delivery-airline.html
15. The Most Amazing Examples of Drones In Use Today, *Forbes:* www.forbes.com/sites/bernardmarr/2019/07/01/the-most-amazing-examples-of-drones-in-use-today-from-scary-to-incredibly-helpful/#5588815f762a
16. Flying High – How a French farming cooperative used drones to boost its members' crop yields: www.sensefly.com/app/uploads/2017/11/flying_high_how_french_farming_cooperative_used_drones_boost_members_crop_yields.pdf
17. How UAVs Are Being Used in Construction Projects: www.thebalancesmb.com/how-drones-could-change-the-construction-industry-845041
18. Airbus launches advanced indoor inspection drone to reduce aircraft inspection times and enhance report quality: www.airbus.com/newsroom/press-releases/en/2018/04/airbus-launches-advanced-indoor-inspection-drone-to-reduce-aircr.html
19. 6 Amazing Passenger Drone Projects Everyone Should Know About, *Forbes:* www.forbes.com/sites/bernardmarr/2018/03/26/6-amazing-passenger-drone-projects-everyone-should-know-about/#785378924 ceb
20. Uber's aerial taxi play: https://techcrunch.com/2018/05/09/ubers-aerial-taxi-play/
21. Airbus's Flying Taxi Is Poised for Takeoff Within Weeks, *Bloomberg:* www.bloomberg.com/news/articles/2019-01-23/airbus-s-flying-taxi-is-poised-for-takeoff-within-weeks
22. China could be the first in the world to start regular flights on pilotless passenger drones: www.cnbc.com/2019/08/28/chinas-ehang-testing-flights-on-autonomous-passenger-drones.html
23. Autonomous weapons: An open letter from AI & robotics researchers: https://futureoflife.org/open-letter-autonomous-weapons/
24. Pentagon seeks "ethicist" to oversee military AI, *The Guardian:* www.theguardian.com/us-news/2019/sep/07/pentagon-military-artificial-intelligence-ethicist#

25. Warning Over Terrorist Attacks Using Drones Given By EU Security Chief, *Forbes*: www.forbes.com/sites/zakdoffman/2019/08/04/europes-security-chief-issues-dire-warning-on-terrorist-threat-from-drones/#e740d287ae41

26. Drone noise is driving people crazy: www.engadget.com/2017/07/18/study-says-drone-noise-more-annoying-than-any-car/?guccounter=1&guce_referrer=aHR0cHM6Ly93d3cuZ29vZ2xlLmNvbS88&guce_referrer_sig=AQAAADMydvLnpdwEE-9CP-wKBhn0Km8EioWM-PDoHDvpcJxMNMkiPJSUZ8MQPkCNp07cDNIOVK_e6olHrY4vStjo I1rCcGowj6eGL8KDh2cLHv7XLcM7aWgFvZOKYU8sstc5STCrE66X rnP8cSxMxW1zVeuAnziWOUDxbQQ-_HvzrGKh

TREND 20
CYBERSECURITY AND CYBER RESILIENCE

The One-Sentence Definition

Cybersecurity is a company's ability to avoid the increasing threat from cybercrime, such as cyberattacks or data theft; cyber resilience is a company's ability to mitigate damage and carry on once systems or data have been compromised.

What Is Cybersecurity and Resilience?

Technology brings unprecedented new opportunities but also unprecedented new threats. In our always-on, always-connected world we see increasing flows of data between companies, their customers, their partners, and their service providers. Businesses rely on interconnected systems and customers expect a 24/7 service.

This means cyber threats such as hacking, phishing, ransomware, and distributed denial-of-service (DDoS) attacks have the potential to cause enormous problems. Services can be disrupted, leading to a loss of trust between customers and providers. Worse still, sensitive personal and financial data can be lost or stolen. Not only is this a blow to consumer confidence that many organizations simply won't be able to recover from, but it can also lead to huge fines from regulators.

A report by IBM found the average cost of a breach to an organization in which personal data was stolen is $3.86 million – equating to a cost of around $148 per stolen record.[1]

Huge fines such as the £183 million British Airways was ordered to pay after customer data was compromised during a cyberattack are making news headlines with increasing frequency. The situation is only likely to get worse as the value of data stored by businesses increases, and hackers develop new tools and techniques involving technology such as artificial intelligence (AI) (see Trend 1). Security specialist 4IQ, which monitors data breaches worldwide, reported a 424% increase in stolen records in circulation during 2018 compared to the previous year.[2]

The loss of customer confidence combined with crippling financial penalties could easily be enough to sink many smaller and medium-sized organizations, and even large corporations can find themselves struggling for many years to fix a damaged reputation. Cyber resilience as a trend is about investing in tools and strategies to overcome these challenges, and ensuring a continuity of service no matter what the online world might throw at you.

Cybersecurity versus Cyber Resilience

So, what is the difference between cybersecurity and cyber resilience? The simple way to think about the difference is that cybersecurity is about stopping threats before they cause damage. Cyber resilience, on the other hand, is about mitigating the potential damage that can be done when your security measures fail.

Cybersecurity is certainly a very important element of any organization, but resilience should involve considering every way in which security failures could impact or decrease the efficiency of business processes.

As no defenses can be guaranteed to be 100% hacker (or mistake) proof, organizations require procedures, tools, and strategies when things do go wrong.

Will any damage, along with regulatory penalties, bring the business grinding to a halt, unable to carry out its basic functions and provide services to customers? Or will well-planned processes spring into life to minimize damage, shore up breaches, and reassure customers that they haven't been put at risk or in danger?

In the past, cyber resilience may have been considered to be solely in the domain of a company's IT department – which would be responsible for installing firewalls, spam filters, and anti-malware measures.

Today, with far more of a business's processes online and reliant on the processing of digital information, the threats are much more diverse and attacks can come from any direction. It is increasingly becoming essential that every department plays its part in upholding the overall level of cyber resilience across the company, and that every member of staff is drilled in recognizing and reacting to the threat of attack. On top of that, companies increasingly rely on cloud and edge computing (see Trend 7), which means cyber resilience has to be considered beyond the boundaries of your company.

How Is Cybersecurity and Resilience Used in Practice?

Whereas cybersecurity is generally used to protect against the threat of attack, cyber resilience measures how efficiently a business can mitigate against the damage caused to its processes (and reputation). It covers threats that can be either adversarial (hackers, thieves, and other malicious actors), non-adversarial (simple human error or incompetence), or both.

One way of thinking about the difference is that cyber resilience involves acceptance of the fact that no cybersecurity solution is perfect or capable of protecting against every possible form of cyber threat. As well as a cybersecurity strategy to minimize the risk of attacks getting through, a cyber resilience strategy is also necessary, to minimize their impact when they inevitably do.

Deepfake Attacks

This involves developing an awareness of the increasingly sophisticated methods being used in cyberattacks. One new technique involves using "deepfake" AI technology (see also Trends 1 and 11) to synthesize the voices of senior leaders within an organization and then make faked phone calls authorizing the release of data or even hard cash.[3] Building awareness of these potential new forms of attack, and understanding how they might affect an organization's capabilities, is an important element of cyber resilience.

Ransomware

Another increasingly common form of attack is ransomware – where attackers encrypt personal or business files and then demand a ransom to unlock them, usually to be paid in anonymous cryptocurrency.

- One of the most infamous examples was the May 2019 **Baltimore** ransomware attack, where municipal computer systems were taken down by hackers who demanded around $72,000 in cryptocurrency to restore access. Even more recently, 23 local government agencies in the state of Texas were taken offline by a similar attack.[4] These attacks brought into sharp focus the fact that compromising essential services can put human lives at risk, and contingency plans are essential in order to provide resilience.

- Projects exist such as the **No More Ransomware** initiative – a collaboration between 36 national law enforcement agencies across the globe, which has so far saved 200,000 victims from having to pay out a combined $108 million[5] – and familiarity with these resources can assist in building resilience. By using this pooled expertise on decrypting files, organizations are able to continue to access their data and carry out their work.

Social Media Hacks

Steps should also be taken to secure social media and other public-facing channels of communication from unauthorized access. Often these are breached by simple social engineering methods, such as tricking authorized users into disclosing their login credentials.

- One recent attack saw hackers take control of the **Metropolitan Police's Twitter and email accounts.**[6] Security breaches like these can lead to loss of trust in frontline channels of communication between organizations and customers, but a cyber resilience strategy can focus on restoring that trust. This could be as simple as having a strategy in place for explaining what went wrong, in an open and honest way, and detailing the steps taken to ensure it won't happen again.

Storing and Protecting Sensitive Data

Good cybersecurity and cyber resilience also means only storing sensitive data when there is a need for it – avoiding the instinct of hoarding data because it might be useful one day, when all it will ever really be is a tempting target for thieves, is a sensible policy when it comes to building resilience.

While hacks and data theft make the headlines, it's important to remember that the term also covers an organization's ability to react and keep functioning in the face of non-adversarial threats, such as accidents, human error, or natural disasters. Any of these clearly have the potential to impact digital operations in a way that can be damaging to a company. Preparing for the damage that these types of incident can cause, and creating strategies for minimizing their impact, should be covered by anyone who is serious about improving their cyber resilience.

Key Challenges

One problem that's likely to be encountered is overcoming the institutionalized beliefs that security and resilience are exclusively the domain of the IT department. Too often, when working with companies on cyber resilience, I've come across a general belief that as long as "the IT guys" do their job, others don't have to worry about hackers, viruses, malware, and system failures.

As I've explained, cyber resilience is a measure of the robustness of an entire organization – not simply its digital systems – in the face of adversity.

These days, even seemingly secure and sophisticated defenses, including firewalls, two-factor logins, and up-to-date anti-malware suites, can be thwarted by careless or unwary individuals installing unsafe software or clicking a dangerous email link.

Certainly, if you're not running the latest, patched, and updated version of all the key tools (including the operating systems) then you're leaving openings that can potentially be exploited.

Understanding how different breaches and failures – from ransomware to accidentally leaving unencrypted customer data on a train – would affect your key operations is essential to planning where resilience-building resources should be placed.

Of course, it can be very difficult to mitigate entirely against the impact of human error. A scary example comes from a web-hosting company, where the owner accidentally removed all traces of his company – together with the online presence of over 1,000 companies that were his clients – by running a destructive code on the company's computers.[7]

Although he later claimed it had been a stunt aimed at generating publicity (not the kind I would want for my own business) it demonstrates how care must be taken to protect against stupidity from inside an organization, as well as maliciousness from outside.

Cyber Resilience and the IoT

The Internet of Things (Trend 2) brings with it a whole heap of challenges when it comes to cyber resilience. In the past, a system failure or hack might only affect the computers you use in your day-to-day work. Today, connected devices and machinery across manufacturing, sales, customer services, R&D, and logistics must be able to run smoothly when events take a turn for the worse.

It has often been found[8] that the "smart" devices – everything from industrial machinery to toys and kitchen appliances that are increasingly ubiquitous in our lives – are often very lacking in even basic security features. This is because they generally rely on security updates and patches provided by the manufacturer, rather than allowing independent measures (such as virus checkers, or anti-malware) to be installed by the user.

Assessing the impact that losing access to the data collected by sensors, cameras, and other IoT-enabled smart devices can have, even in the short term, is essential. For example, if you use computer vision (Trend 12) cameras as part of your QA processes, those processes could come to a grinding halt if the cameras are brought down by a DDoS attack.

This increasingly connected environment means that cyberattacks and failures that previously would only affect one isolated resource or process could potentially spread to any area of your operations. For this reason, if you rely on connected devices to carry out services your business and customers rely on, IoT can prove a serious challenge when it comes to improving cyber resilience.

The key principle of cyber resilience is ensuring continuity. Identifying areas where interruptions are likely to cause problems and ensuring processes are in place to keep the cogs turning (and customers smiling) is central to understanding the importance of this trend.

How to Prepare for This Trend

Cybersecurity is undoubtedly the first line of attack. Ensuring all of your devices are running on the most up-to-date firmware, that firewalls, VPNs, and anti-virus/malware software is running, and all of your software and tools are fixed with the latest patches is an important first step.

To improve cyber resilience, you can start by working to identify where events and incidents could have the most damaging effects. Drawing up a list of where your operations are reliant on technology, as well as where sensitive and valuable data is stored and used, will help you to gain an overall understanding of how continuity of service could be affected.

This is where the concept of a "digital twin" (Trend 9) can play an important role in cyber resilience. A digital, simulated model of your organization or its processes can help to understand the impact on overall output and efficiency that adverse incidents can have.

Once there is an understanding of how core functions could be affected, measures can be put in place to mitigate as best as possible in the event of attack of failure. This could involve development of offline emergency processes for keeping essential functions such as QA, finance, customer service, and security running as optimally as possible, until the breach can be fixed and normal services resumed.

Another crucial step is the creation of a solid response plan, clarifying not only what needs to be done in the event of a breach or failure, but who is responsible for doing it, how failures should be reported, and how their impact should be measured. One way of doing this is by putting together a response team, with representatives in every business department, who are responsible for declaring a "state of emergency," coordinating remedial actions, and reporting the success or failure of resilience measures within their area of responsibility.

In many businesses, customer services will have a core role to play in the event of service disruption caused by cyber adversity. Reassuring customers that they can continue to rely on you, and that their data is safe, will be an important consideration – as loss of trust is one of the most serious dangers to continuity that cyber resilience is designed to mitigate against.

The final step of a resilience plan is recovery – this means resuming normal operations as quickly as possible as well as restoring data that may have been lost or isolated. If data has been lost or accidentally erased, this could be a lengthy process – unless a comprehensive policy of backing up is part of your wider resilience strategy. Cyber resilience promotes the idea that it's impossible to certify that any piece of data within any organization is 100% "safe," and therefore steps should be in place to recover it when it is lost. Understanding who is responsible for recovering data and returning processes to normal operations, and the steps which must be taken to achieve this, is the last piece of the puzzle.

The steps outlined above are based on the cybersecurity framework developed by the National Institute of Standards and Technology[9] – although each one is expanded to consider the impact that adverse cyber events could have on continuity of operations, as well as the immediate threats they pose.

In order to achieve all of this, it will usually be necessary to educate everyone in the organization – from the C-level executives to shop floor workers – on the principles of both security and resilience. In a large organization this is certainly no mean feat – but the investment in time and resources here will be repaid many times over once you've withstood your first cyberattack or the impact of your first slip-up.

Notes

1. 2019 Cost of a Data Breach: www.ibm.com/security/data-breach
2. 4IQ Identity Breach Report 2019: https://4iq.com/2019-identity-breach-report/
3. Fake voices help cyber crooks steal cash: www.bbc.co.uk/news/technology-48908736
4. Texas government organizations hit by ransomware attack: www.bbc.com/news/technology-49393479
5. The quiet scheme saving thousands from ransomware: www.bbc.co.uk/news/technology-49096991
6. Met Police hacked with bizarre tweets and emails posted: www.bbc.co.uk/news/uk-england-london-49054332
7. Man accidentally "deleted his entire company" with one line of bad code, *Independent*: www.independent.co.uk/life-style/gadgets-and-tech/news/man-accidentally-deletes-his-entire-company-with-one-line-of-bad-code-a6984256.html
8. IoT Security is being seriously neglected: www.aberdeen.com/techpro-essentials/iot-device-security-seriously-neglected/
9. What is the CIF?: www.nist.gov/itliadindex/visualization-and-usability-group/what-cif

TREND 21
QUANTUM COMPUTING

The One-Sentence Definition

The cheap smart phones we carry in our pockets today are thousands of times more powerful than the computers used to put humans on the moon just half a century ago, while the arrival of quantum computing will make today's state-of-the-art look Stone-Aged.

What Is Quantum Computing?

Traditional computers may have grown exponentially more powerful over the last century, but at their heart they are still just very fast versions of the simplest electronic calculators. They are only capable of processing one "bit" of information at a time, in the form of a binary 1 or 0.

Quantum computing harnesses the peculiar phenomenon observed to take place when operating at a subatomic level – such as quantum entanglement, quantum tunneling, and the ability of quantum particles to apparently simultaneously exist in more than one state. Using these methods, it has been demonstrated that it is possible to build machines capable of operating far more quickly than the fastest processors available today – potentially hundreds of millions of times faster, in fact.[1]

Researchers at Google – who announced this year to have completed the world's first calculation using quantum computers that would be impossible to complete on non-quantum machines – have said that the development will lead to the shattering of Moore's Law.[2] Moore's Law was defined in 1965 and states that computing power will double roughly every two years.

As an example of this growth in power, consider the 2048-bit RSA encryption algorithm, currently used to secure internet traffic as well as the most sensitive data transmitted by corporations and governments. At the end of the day, it is just that – an algorithm – and just as it must be capable of being decoded using a key, it can also be decoded through brute force (trying every number until the right key is found).

Currently, it's thought that it would take commonly available computers millions of years to brute force or "guess" their way into a piece of information encrypted using 2048-bit RSA. However, it was recently demonstrated that quantum computers could be capable of the feat in around eight hours.[3]

Rather than the standard binary bits, quantum computers use "quantum bits," known as qubits, to process data. Key to their power is that qubits seem to be capable of existing in states of both 1 and 0 at the same time. The most powerful commercially available quantum computers in existence today generate around 50 qubits,[4] while the hypothetical machine required to crack 2048-bit RSA in eight hours will require around 20 million. If that seems like a huge jump, then consider that the power of the most advanced computers available has already increased one trillionfold in the last 60 years.[5]

Beyond that, there are of course hugely positive implications for this literal quantum leap in the amount of processing power available to us. While conventional binary computing is likely to be all we will need for many of the tasks we carry out on computers in the near future, incomprehensibly quick quantum computing is likely to have a variety of applications in fields such as artificial intelligence (Trend 1) and decoding complex structures such as genomic information (Trend 16).

Most experts agree that truly useful quantum computing is not likely to be a feature of everyday life for quite some time. However, big players in tech, including Google, IBM, and Intel, have already made platforms available that enable anyone to experiment and benefit from these early forays into subatomic architecture. And even putting quantum power aside, the power of traditional computer processors will undoubtedly continue to increase at breakneck speed, just as it has done throughout the last half century.

Besides quantum computing, other technologies are under development that could allow us to build far more powerful machines in the near future. One development, known as PAXEL, uses "nanophotonics" to carry out calculations using varying intensities of light level. This speeds up the rate at which data can move across integrated circuit boards, offering faster processing speeds and lower energy consumption.[6]

Just as ever-increasing networking speed (see Trend 15) means more than simply faster data transfer, also unlocking new possibilities for applications that previously would have been impossible, increasing processing power broadens the horizons of what technology can do for us.

How Is Quantum Computing Used in Practice?

Current applications of quantum computing can seem impossibly Space Age and "out there" – this is because mostly it is confined to academia and highly theoretical work, with little in the way of practical applications.

- Building time crystals, for example – solid structures that contain molecular patterns that alter in a predictable manner throughout time, as well as space (as with regular crystals) – is an interesting intellectual challenge for a theoretical physicist but of very little use to most of us. And it's currently only possible using quantum computing.[7]

- The first company to commercially exploit quantum computing was **D-Wave**, based in Canada, which provides services to Lockheed Martin and NASA, among others. Back in 2012, its D-Wave One system was used by **Harvard University** researchers to tackle the problem of protein folding[8] – predicting the physical form a protein will take when it acquires its three-dimensional form – a necessary step on its path to becoming biologically functioning. "Misfolded" proteins in living organisms can result in diseases and allergies, so understanding this process is highly valuable in the field of medicine. Predicting the results of protein folding is possible using conventional computing (as seen in the crowd-sourced, distributed folding@home project[9]) but quantum computing has the potential to speed up the process while enormously reducing the amount of energy required.

- One project using D-Wave technology was carried out by researchers at **Volkswagen**, who used quantum methods to model the flow of traffic through city centers[10] – an enormously complex task that is very hit-and-miss using today's technology. This is hardly surprising considering that introducing just 270 on-or-off variables (such as whether a particular car travels onto a particular street) into a simulation results in more potential outcomes than there are atoms in the universe!

- In fact, quantum computing offers the potential for carrying out just about any form of complex modeling with far greater efficiency. Weather forecasting, for example, relies on our ability to model complex meteorological systems. The **UK's Met Office** has said that it considers quantum computers offer the potential for carrying out far more advanced modeling than is currently possible today, and it is one of the avenues being explored for building next-generation forecasting systems.[11]

- Along with protein folding and meteorology, systems don't come much more complex than **global financial markets**. Fully understanding the movements of stock markets, currencies, and commodity prices is currently beyond the scope of even the most sophisticated machine learning algorithms running on classical computing architecture. Prices are influenced by global politics, local economics, consumer trends, scientific advances, social shifts, wars, natural disasters, and things posted on Twitter by celebrities. It may never be possible to fully calculate the impact of all of this chaos on structured, man made systems like markets. But massively increasing the processing resources available for interpreting data and building simulated models is certain to lead to better predictions.

- As well as developing superior artificial intelligence capable of predicting and reacting to the movements of fickle financial markets, it could also greatly improve our ability to recognize and tackle fraudulent activity among the thousands of transactions processed by large banks and payment providers every second of every day. It could also allow for far more reliable, and fair, **credit scoring**.[12]

As quantum computing is thought to hold huge potential when it comes to developing new pattern recognition and optimization methods, a great deal of money could be made by whoever is first to successfully apply these novel computing methods in the world of finance, or in other complex and lucrative areas.

Key Challenges

One of the biggest challenges posed by the emergence of these new classes of computer processor is that they will often require software that has been specifically programmed to run on them. It's highly unlikely that you will simply be able to take a quantum or nanophotonic-powered CPU, plug it into your laptop, and expect it to turbocharge your Windows PC!

In classical computing in recent years, we've seen the trend of ever-increasing CPU clock speeds begin to take a back seat to multi-core architectures. Packing more cores into processors allows them to carry out more tasks simultaneously, which can lead to huge performance increases – but only when running software that has been specially programmed for it.

The appearance of exotic new data processing technology will require new software architecture before we are able to create tools and applications that will take full advantage of them. For software engineers, this may mean going back to basics to learn a whole new skill set, as well as educating themselves with a thorough grounding in quantum physics if they really want to be able to understand what they are doing.

Until such time as it's commercially worthwhile for large numbers of smart people to spend their time learning these skills, it may be difficult to find people with the ability to work on projects involving quantum and other advanced forms of computer processing.

Which brings us on to another challenge for anyone wanting to cash in on quantum – the lack of obvious commercial opportunities. Until very recently, when Google announced that it had achieved the first instance of "quantum supremacy"[13] – conducting an operation on a quantum machine that wouldn't be possible on a classical computer – it couldn't do anything that couldn't be done more cheaply by other means.

This means that there's a real need for balancing the amount of time and resources that goes into research into possible future uses with those spent on meeting today's needs. While it's great to know that you'll be able to predict every possible move and fluctuation of your industry in 20 years' time, spending all of your resources on trying to do it now is likely to lead to a loss of competitiveness in the short term. Ignoring it, however, leaves you facing obsolescence when someone else eventually works out how to capitalize on its potential.

Cloud-powered quantum-as-a-service is already available from Google, Microsoft, and IBM, meaning the power is there for those that need it. But evaluating what you have to gain while avoiding the trap of "fear of missing out" is important.

If you're looking to explore beyond what is currently possible with the roughly 50 qubits of power offered by these services, there's the problem that on a technical level, quantum (and other advanced models of computing) is extremely expensive and tricky.

Quantum computing can only take place in extremely cold conditions – the inside of quantum machines such as D-Wave 2X operates at a temperature of 15 millikelvin – a fraction of a degree away from absolute zero and 180 times colder than the temperature in interstellar space.[14] This is because the subatomic particles must be as close as possible to a stationary state in order to be measured. Only the world's most advanced and well-funded organizations and institutions are capable of doing this at the moment. However, this is likely to change as understanding of the technology increases and more commercial possibilities become apparent.

How to Prepare for This Trend

Working out where an increase in processing power is likely to open up new possibilities for innovation and leadership, rather than just "more of the same, but faster," is essential if you want to harness the potential of this trend.

In truth, quantum computing is unlikely to start having a hugely noticeable effect on our everyday lives for quite some time. However, even though truly useful everyday applications of quantum computing are considered to be at least a decade away, it is expected to be so revolutionary that developers and engineers with an eye on the future are starting to prepare for its impact now.

If you are in an industry that involves simulation, modeling and prediction of complex, chaotic systems, such as finance, pharmaceuticals, or IT security, then the day its impact begins to be felt may not be too far away.

Cryptography is one field where this is apparent. Already a great deal of work is going into the development of quantum-safe algorithms so that secure data transmission and processing will still be possible when the quantum age arrives.

While this may not seem like a huge issue to most of us – who are unlikely to care too much if the confidential information we exchange online, such as our credit card numbers, are decoded in 20 years' time – for governments and organizations today this already poses a real security threat. There may well be a need for information sent today to be kept confidential beyond that relatively short span of time.

Working to understand the limitations of existing classical systems – the point at which the systems you are working with start to become too complicated to predict – can help a business or organization to see where quantum or other advanced processing models may one day be a good fit or even become a necessity.

For the more technically minded, a number of quantum computing programming languages and software development kits already exist, as well as a good number of online resources which explain how they work and how to use them. Studying these would be a good step towards understanding what sort of problems quantum computing is likely to help us solve. These include Ocean, developed by D-Wave, Cirq from Google, Q Sharp from Microsoft, and Qiskit from IBM. All of these are open source projects that allow experiments to be carried out with code that can either be run locally on quantum simulators, or sent to the cloud (Trend 7) for processing with actual quantum devices.

Increasingly, the fundamentals of quantum computing are becoming course components of graduate and even undergraduate maths, physics, and computer science degrees. A thorough academic grounding in these subjects is likely to be a good foundation for anyone wanting to go on to study the theoretical and practical applications of this technology trend – and develop a skill set that is likely to be in high demand in the very near future.

Notes

1. Where do quantum computers get their speed: http://quantumly. com/m.quantum-computer-speed.html
2. Google claims to have reached quantum supremacy, *Financial Times*: www.ft.com/content/b9bb4e54-dbc1-11e9-8f9b-77216ebe1f17
3. How a quantum computer could break 2048-bit RSA encryption in 8 hours: www.technologyreview.com/s/613596/how-a-quantum-computer-could-break-2048-bit-rsa-encryption-in-8-hours/
4. Google's "Quantum Supremacy" Isn't the End of Encryption, *Wired*: www.wired.com/story/googles-quantum-supremacy-isnt-end-encryption/

5. Visualizing the Trillion-Fold Increase in Computing Power: www.visual capitalist.com/visualizing-trillion-fold-increase-computing-power/
6. Using light to speed up computation: www.sciencedaily.com/releases/ 2019/09/190924125018.htm
7. A team of University of Maryland researchers have developed the world's first time crystals: https://dbknews.com/2017/03/17/time-crystals-discovery/
8. D-wave-quantum-computer-solves-protein-folding-problem.html: http://blogs.nature.com/news/2012/08/d-wave-quantum-computer-solves-protein-folding-problem.html
9. Folding@Home – About: https://foldingathome.org/about/
10. Traffic Flow Optimization using the D-Wave Quantum Annealer: www. dwavesys.com/sites/default/files/VW.pdf
11. Novel architectures on the far horizon for weather prediction: www. nextplatform.com/2016/06/28/novel-architectures-far-horizon-weather -prediction/
12. Quantum computing for finance: Overview and prospects: www.science direct.com/science/article/pii/S2405428318300571#sec0009
13. Google quantum computer leaves old-school supercomputers in the dust: www.cnet.com/news/google-quantum-computer-leaves-old-school-supercomputer-in-dust/
14. The D-Wave 2X Quantum Computer Technology Overview: www. dwavesys.com/sites/default/files/D-Wave%202X%20Tech%20Collateral _0915F.pdf

TREND 22
ROBOTIC PROCESS AUTOMATION

The One-Sentence Definition

Robotic process automation (RPA) is technology that can automate business processes that are rules based, structured, and repetitive.

What Is Robotic Process Automation?

RPA is about minimizing the time spent on routine, manual activities that are currently performed by humans but could be given to software robots. The aim of RPA is to improve productivity, reduce human error rates, and free people up to do the higher value work that can't yet be performed by robots, such as solving more complex customer queries or developing "big picture" strategies. Consider RPA "robots" that are programmed to complete specific business processes such as interacting with other digital systems, capturing data, retrieving information, processing a transaction, and more.

Just as physical robots (see Trend 13) have been developed to carry out many elements of physical manual work, such as conveyor-belt manufacturing processes, or even building houses,[1] software robots can be used to carry out repetitive digital tasks. This generally requires the software to be "taught" how to do the job. However, increasingly, advanced implementations will make use of artificial intelligence (Trend 1) to become increasingly proficient – and eventually to automate tasks themselves, or prioritize which tasks should be automated.

RPA shouldn't really be thought of as new processes, but instead as a new "layer" of activity that sits above all of your processes that involve manual or repetitive tasks. They operate independently, and if for some reason they fail, humans can carry on getting the job done, albeit in our own slower, more error-prone manner. Just as how in the house-building example all of the raw ingredients of the house start and end up in the same place whether the work is carried out by a robot or a human, the same is true of software RPA. It's just the middle-man work of actually moving the bricks – or in the case of software RPA, data – from place to place, which becomes automated.

In other words, they don't replace existing systems or applications, but complement them and act as a buffer between them and a human workforce.

We saw the beginnings of RPA in functions such as auto-complete used in web forms and productivity tools such as macros in spreadsheets, as well as email auto-responders. Today they have evolved into customer service chatbots – used to deal with less complex customer queries.

Until now, RPA deployments generally involve feeding machines a structured set of rules. However, as RPA technology evolves, cognitive computing technologies such as computer vision (see Trend 12) and natural language processing (see Trend 10) will broaden their capabilities and make it simpler to automate an ever-increasing number of tasks. Automatically reading in data from hand-filled forms, or from footage captured on cameras and other sensors in an industrial environment, will unlock countless opportunities. RPA tools of the future will simply be watching you to identify tasks they could help with. Just imagine an intelligent RPA tool watching you send emails. I find myself drafting very similar email responses to similar requests, such as consulting or speaking engagements. Even the follow-up conversations are very similar in nature. There is nothing stopping RPA tools suggesting or drafting those emails for me after they have learned from how I would normally respond.

In fact, analysts at Gartner have predicted that 85% of large organizations will have deployed some form of RPA by 2022,[2] while Forester researchers state that spending on RPA tools will grow by a third, from around $1 billion in 2019 to $1.5 billion in 2020.[3]

As they generally require little change in physical infrastructure, RPA procedures can be relatively cheap to implement, and the benefits of freeing up huge amounts of staff time and reducing error rates can quickly outweigh the expense. This is why RPA is often considered a quick win – benefits quickly become apparent and this can be used to win buy-in for further projects involving technology-driven change.

How Is Robotic Process Automation Used in Practice?

RPA is sometimes referred to as "white collar automation" because it is used to automate processes carried out by clerical, administrative, managerial, and professional personnel in many industries.

RPA systems are built into much of the software and online systems that we use in everyday life, such as spellcheckers, auto-completing forms, and password lockers. In fact, as I am writing this, Microsoft Word is automating the process of keeping my endnotes in tidy, numerical order. This philosophy has carried over into business processes, with the financial services in particular quickly finding ways to put it to use.

RPA in Financial Services Firms

Banks and insurance companies have embraced RPA; let's look as some examples.

- **American Fidelity Insurance** says it has managed to save 10 hours of human work for every hour they have put into building robots that are used for both claims processing and accounting tasks. It also uses machine learning to infer the correct destination for an email based on its contents, and route it to the right recipient – a function previously carried out by humans.[4]

- Singapore bank **OCBC** says it has cut down the time it takes to reprice a home loan from 45 minutes to one minute by automating processes involving checking eligibility, recommending options to customers, and drafting emails.[5]

- Meanwhile, **DBS Bank** partnered with IBM to develop its Centre of Excellence for RPA, which was used to optimize more than 50 business processes.[6] IBM Managing Director for Banking Adam Lawrence said, "While the ability to automate processes has existed for a very long time, technology has since evolved to enable cognitive automation through autonomous decision-making, new insights through data discovery, and personalized support."

- Another major bank turned to RPA to speed up a compliance process that involved employees needing to monitor more than 200 websites to keep up to date on changes to rules and regulations. Using solutions provided by RPA vendor Kryon, they were able to automate logging into these sites and gathering the required information, reducing the time spent on this task by humans from one to two hours, to 20 minutes, while also reducing the rate of errors.[7]

RPA in Retail

- In retail, **Walmart** uses around 500 bots to carry out tasks involving everything from customer service, employee relations, auditing, and paying invoices. According to CIO Clay Johnson, "A lot of those [ideas] came from people who are tired of work."[8]

- And Australian wholesaler **Metcash** has implemented a system that involves a committee looking for "low-hanging fruit" to spot opportunities for rolling out RPA. Program manager Jennifer Mitchell explained in an interview that "RPA is not a Band-Aid for a poor process. Our mantra is to give the human back the human, and the way we do that is to look at any process that is digitized and how it can be streamlined and automated."[9] So far, the company has identified 20 opportunities for implementing RPA into low-value, repetitive tasks, and the plan is to have 30 in place in 2020.

RPA in Healthcare

Of course, healthcare is another industry that necessarily involves a lot of form-filling and back-office administrative work. Patient records need to be kept with meticulous accuracy, and made available when they are needed, while maintaining the need for privacy. RPA systems allow data to be readily extracted from handwritten notes, charts, medical images, and observational data, meaning clinicians will always have the most up-to-date information about a patient at their fingertips. They can also encode information with metadata, meaning personal details don't have to be divulged to every system that will read the data.

This functionality can also be extended to patients, who can check in, confirm their identity, and provide their consent through digital interfaces. This data can then be used to automatically populate and authorize forms and records.

RPA in Customer Services

Customer services also presents many opportunities. Chatbots are becoming increasingly common, and thanks to advances in natural language processing are constantly improving their ability to understand and assist customers. This frees up the human customer services operator from repeating the same advice over and over again, meaning they can spend more time on the complex cases that require human intervention.

- A good example is the insurance call center where RPA was implemented to allow call handlers to carry out compliance checks while they were on a call. Rather than spending time gathering the necessary information from various sources while the customer waited, automated processes were able to gather the information in the background and update customer records, resulting in a cut in the average call time of 70% and reducing the amount of time spent by customers waiting on the line from two minutes to 40 seconds per call.[10]

- Tools are now emerging that are capable of analyzing processes and systems in order to determine where RPA can be rolled out automatically. Cloud-based HR specialists **PeopleDoc** use a system involving "PeopleBots" that effectively monitor the way business operations work, and look out for situations where automation would be helpful.[11] When they are found, machine learning algorithms determine the best way to automate the process and start getting the job done.

Key Challenges

Probably the most important challenge is the impact that RPA could potentially have on people's jobs. While it's certainly possible that it will make humans redundant when it comes to a large number of tasks, new tasks will be required of people who are capable of building and deploying automated processes.

Forrester research suggests that RPA and other automation processes will replace 16% of US jobs by 2025; however, the equivalent of 9% could be created, leading to a net loss of 7%.

As well as jobs disappearing or entirely new ones being created, many people are likely to find that what is expected of them in their existing roles will change. As the more mundane and routine elements of their day-to-day activities become automated, there will be more time to spend on the creative, strategic, or customer-facing tasks. This could bring about major cultural shifts, such as greater opportunities for working remotely and "in the field," when less time needs to be spent sitting at a desk in front of a computer terminal. While this may sound completely positive – and if managed well, it should be – there may be resistance from those who have become used to routine. It will also mean a need to educate and train staff on how to make the most of their newly available free time.

As well as the social challenges, there are obviously technological hurdles that need to be cleared, too. Planning, building, and deploying RPA is likely to require new skill sets that may need to be hired in, or the existing workforce upskilled. Deploying RPA means building systems over the top of existing ones using new tools, which are unlikely to be in the toolbox of most organizations. Assessing whether the benefits justify the expense will become a highly valued skill – particularly in the face of evidence that shows that many RPA implementations have failed to deliver ROI.[12]

This means that selecting the right tasks to automate is vital and expectations must be kept realistic. While artificial intelligence is undoubtedly expanding the range of situations where RPA is suitable, at the moment it is only viable for repetitive, high-volume tasks. For tasks where human oversight or decision-making is essential, the gains in speed will be lost due to the need for manual intervention, and the reduction in human error will not be guaranteed.

Questions also need to be addressed over whether or not automation is the right solution for a troublesome (i.e. time consuming or boring) process. In some cases it may be the process itself that is at fault – and careful consideration may have to be given to whether it is achieving the intended aims or filling a necessary requirement. If the repetitive nature of a task is requiring too much in the way of resources, applying RPA may simply be "papering over" the problem, rather than solving it. This could be a dangerous and costly mistake, if refining or rethinking the process would be more effective than automating it!

How to Prepare for This Trend

The first step is to understand the basics of what types of process can and can't be automated. Typically, this is the "busy work" which some estimates say takes up between 10% and 20% of the typical information worker's day.[13] Work that involves a great number of repetitive actions – opening and searching records, transferring data between different digital locations, and repetitive mouse clicks – are prime candidates for automation, while jobs that involve creative thought and human decision-making, generally, are not.

Next, you will need to identify which tasks should be automated. These will be ones that help your organization achieve its overall aims, but currently consume a disproportionate amount of employees' time. Remember, it's usually a good idea to go for "quick wins" first – these will help to establish the usefulness of RPA, while winning over minds that may be resistant to the idea of reducing repetitive workloads, or fearful of what it could mean for jobs and organizational culture.

Following this you can begin researching the technologies that are available, and the potential partners you may need to work with to create a successful deployment. Other considerations here will include your existing infrastructure – is it built in a way that will allow RPA to be deployed on top of it? And what will your existing workforce need in order to capitalize fully on the time and opportunities that are unlocked for them?

When picking a partner, look at providers that have a proven track record in your industry, and those that can help you manage the human element of the incoming change – as this will undoubtedly be the most difficult element of the automation process to predict and deal with.

At Adobe, the path to RPA began with a "quick win" proof of concept, followed by a number of pilot programs designed to integrate RPA into finance tasks. When the benefits became apparent, the next step was to establish an RPA "center of excellence" in order to take a holistic view of when and where automation would be appropriate, and to aid in the development of a reusable set of tools and skills.[14]

As with everything related to technological change, there's a great deal of information and resources available online that you can use to educate yourself and prepare for change. RPA solutions provider UIPath have created an online academy where a number of free training courses are available. They are categorized according to the roles that need to be filled in order to deploy RPA – such as business analyst, implementation manager, and solution architect – with each one covering a specific step of the journey towards RPA.

As well as this useful resource, UIPath has built a community of RPA developers and enthusiasts to share tools, tips, and strategies.[15] Through it, you can experiment with free, customizable robots designed to help organizations understand how RPA works and where it can be useful.

Notes

1. The House The Robots Built: www.bbc.com/future/bespoke/the-disruptors/the-house-the-robots-built/

2. Gartner Says Worldwide Spending on Robotic Process Automation Software to Reach $680 Million in 2018: www.gartner.com/en/newsroom/press-releases/2018-11-13-gartner-says-worldwide-spending-on-robotic-process-automation-software-to-reach-680-million-in-2018

3. How Automation Is Impacting Enterprises In 2019: https://go.forrester.com/blogs/predictions-2019-automation-technology/

4. RPA is poised for a big business break-out: www.cio.com/article/3269442/rpa-is-poised-for-a-big-business-break-out.html

5. Examples and use cases of robotic process automation (RPA) in banking: www.businessinsider.com/rpa-banking-examples-use-cases?r=US&IR=T

6. DBS Bank accelerates digitalization transformation with robotics programme: www.dbs.com/newsroom/DBS_Bank_accelerates_digitalisation_transformation_with_robotics_programme

7. RPA Use Cases: www.kryonsystems.com/Documents/Kryon-UseCases-Financial.pdf

8. What is RPA? A revolution in business process automation: www.cio.com/article/3236451/what-is-rpa-robotic-process-automation-explained.html

9. What RPA Really Means for Managing Accounting: www.intheblack.com/articles/2019/11/01/what-rpa-really-means-for-management-accounting

10. RPA Use Cases in Call Centers: www.kryonsystems.com/Documents/Kryon-Call-Center-Use-Cases.pdf

11. Robotic Process Automation and Artificial Intelligence in HR and Business Support – It's Coming: www.bernardmarr.com/default.asp?contentID=1507

12. Why RPA Implementations Fail: www.cio.com/article/3226387/why-rpa-implementations-fail.html

13. Robotic Process Automation: Statistics, business impact and future: www.pavantestingtools.com/2017/10/robotic-process-automation-statistics.html#.WsaCtdPwbMI

14. Adobe CIO: How we scaled RPA with a Center of Excellence: https://enterprisersproject.com/article/2019/10/rpa-robotic-process-automation-how-build-center-excellence

15. UIPath Go: www.uipath.com/rpa/go

TREND 23
MASS PERSONALIZATION AND MICRO-MOMENTS

The One-Sentence Definition

Mass personalization is about offering products and services at scale, but each uniquely tailored to our needs; micro-moments are the opportunities to respond to customer needs at an exact time when they need them.

What Are Mass Personalization and Micro-Moments?

Mass Personalization

Targeted mass marketing was developed by direct mail businesses in the 1960s and 1970s, to enable customers to be segmented by age, geography, or income and offered products they are more likely to be interested in.

Today, due to the internet, social media, and our always-connected society, more information is generated about who we are and what we do than ever before (see also Trend 4 on big data) – and all of this can be captured and analyzed by marketers and trend spotters.

This means they can offer us products and services uniquely tailored to meet our individual needs in increasingly individual ways. From personalized email marketing to setting tariffs based on our level or volume of consumption, mass personalization is now used to drive higher sales, increased customer satisfaction, and, hopefully, improved retention.

Online, data-driven mass personalization started out with determining a user's geographical location from their IP address and directing them to a landing page serving their particular region. As the variety and volume of the data collected increase, so does the granularity with which customers can be segmented by age, interests, occupation, or many other factors that can be determined. This means that the picture the marketer has of each customer group becomes increasingly personal.

The desire for personalization has become increasingly important to marketers as we consumers have become increasingly resilient towards, and resentful of, poorly targeted mass communications from people trying to sell us things. This is shown in research by Deloitte that states that 69% of us have unfollowed a brand on social media, closed an account, or cancelled a subscription due to annoying or irrelevant advertising.[1] It seems somewhat contradictory then that the same research shows that only one in five people is comfortable with the basic idea of businesses using their personalized information to offer them more relevant products. In spite of this, and in spite of increasing regulations governing how this data can be used, plenty of us wittingly or unwittingly give consent for our activities to be tracked and analyzed – even if it's just because we don't read the terms and conditions properly!

At the same time, interest and demand for personalized products and services is undoubtedly increasing. We're used to being able to customize big-ticket purchases like houses and cars, with manufacturers and builders offering a range of extras and additions to tempt us to increase our spending. And buyers of luxury items such as jewelry and bespoke clothing have always been able to have their own personal flourishes added at the point of production. Today, mass-market products can also be individually customized to help the buyer feel they have spent their money on something uniquely special to them. Often this can involve encouraging the consumer to get personally involved in the customization process, offering them web portals where they can design their finished product and kit it out with all the extras.

Automated retail, robotic manufacturing (Trend 13), and 3D printing (Trend 24) are all technological trends that have made it easier to offer personalized goods and services, and businesses that have successfully built these capabilities have gone on to become market leaders. As Deloitte's report concludes, "In the future, businesses that do not incorporate an element of personalization into their offering risk losing revenue and customer loyalty."

Micro-Moments

Traditionally marketers have looked to capture information about what we do, when we do it, why we do it, where we do it, and who we do it with. This information is then corelated to try and draw up overall pictures of who we are and work out the best way to sell us stuff that we may or may not need.

As technology has become more powerful and clever when it comes to crunching this information, it's becoming possible to do this in close to real time – meaning marketers can approach us closer and closer to the point in our lives when we are looking for a product or service.

An early breakthrough was made when US retailer Target announced that it had worked out how to take an educated guess that someone had become pregnant – possibly even before their family or friends know.[2]

Today, the trend has moved beyond predicting major life-changing episodes like becoming pregnant or engaged, and towards identifying what we are doing moment to moment. As an extreme example, take Facebook's reported strategy of determining when people are likely to be feeling depressed about their appearance and offering them products related to weight loss or improving their looks.[3] Creepy, certainly, and unethical if it's used to exploit people.

But if it's done with care to avoid exploitation and takes an ethical approach to privacy and data collection, this type of marketing has the potential to impact our lives in positive ways by giving us access to products and services at the time we need them, reducing waste generated by poorly targeted advertising, and cutting down on the amount of irrelevant and annoying commercials we have to put up with.

The perfect micro-moment occurs when an offer of a product or service pops up in front of us exactly at the time when we are trying to solve a problem that it might help us with. This might mean deciding what movie to watch next, what to wear to a wedding, or how to make the journey to somewhere we need to go. Predicting when these windows of opportunity are likely to open up means businesses can enter our lives at the right time, greatly increasing the efficiency of their marketing operations.

The term itself is often said to have been created by Google[4] – to describe "intent-rich" moments where marketers can take advantage of the growing expectation of instant gratification that today's always-connected culture has developed in us.

How Are Mass Personalization and Micro-Moments Used in Practice?

The internet giants like **Google**, **Facebook**, **Netflix**, **Amazon**, and **Spotify** have pioneered the trends of personalization and identifying micro-moments by learning to serve up personal recommendations at the time we're likely to want them. When you search for a product on Amazon or scroll through the movies available on **Netflix**, you're being offered a sample of what is available based on what the service provider thinks you will want.

In particular, **Google**, as well as other search engines like **Baidu**, or Microsoft's **Bing**, are known to be increasingly personalizing the results of web searches. This means that as well as the basic metrics, such as how many other pages link to a particular page, that are used to determine the likelihood of information appearing in your results, factors that are personal to you – such as your location, demographics, and search history – are taken also into account.[5]

Taking their cues from this, businesses in just about every industry have jumped on the bandwagon and are increasingly likely to be putting resources into chasing the same results.

In food retail, large amounts of stock often goes unsold and spoils, due to errors in predictions about what customers will want to buy.

- Retailers like **Walmart** and **Tesco** now use highly targeted analytics to understand how much of each type of produce is likely to sell in each store.

- And **Amazon** has talked about predictive shipping that will allow them to dispatch products to customers before they have ordered them – safe in the knowledge that they are likely to be wanted.[6]

- **Coca-Cola** change their marketing around the globe, with advertising and packaging altered to account for local cultures and traditions.

- Innovators such as Indian soft drink brand **Paper Boat** have taken this further. The company adds flavors to its drinks to suit regional tastes, and even uses locally sourced mangos so local consumers will recognize the flavor when they taste it.[7] It gathers data on customer preferences through WhatsApp surveys, and altering the recipe of drinks being created in its automated factories takes "at most two to three minutes."

- Headphone manufacturer **Revols** successfully crowdfunded the production of earphones that are self-molding to permanently shape them to fit their owner's ears within one minute of putting them on.[8] Previously, filling these audiophile needs (or just fitting unusually shaped ears) required custom-fitted earphones and would involve paying thousands of dollars to a specialist audiologist. Innovation like this is a good demonstration of how manufacturers are helping to bring a bespoke level of craftsmanship usually restricted to luxury or big-ticket items to the mass market.

Other innovators are actually offering personalized services.

- A great example is the **Vi** interactive personal training program, which develops a personal running routine based on what it knows about the user's fitness, activity, and performance levels.

- In news publishing, there's a constant race to be the first to provide readers with up-to-date information about the world, which is relevant or of interest to them. Big news organizations have invested heavily in personalized news, serving up stories that they predict their readers will be interested in. Chinese news aggregation app **Toutiao** claims to be able to get an accurate understanding of the news a particular user will be interested in in just 24 hours.[9]

- The beauty industry has also been quick off the mark to capitalize on this trend. German startup **Skinmade** creates customized skin creams using machine learning after analyzing an individual's skin condition at its custom kiosks, where the cream is formulated and dispensed while the customer waits.

* And **Neutrogena** has developed an app that allows users to scan their face with their camera and create custom-fitted and formulated beauty masks which are delivered to their door.[10]

Mass personalization also describes many technology-driven trends that are revolutionizing healthcare. These include personalized DNA testing which can highlight individual risks, targeted gene therapy, and services which can provide personalized reports and diagnoses based on analysis of medical notes and images.[11]

Of course, there have been more sinister applications of technology used when chasing these two particular trends. Of note is the incident involving Facebook, mentioned above. In 2017, leaked documents suggested that the social media giant had the ability to determine the mental state of users, based on how they interacted with the service. This pitch to advertisers was reported to suggest that they could use this information to pinpoint the correct time to target them with advertising for feel-good products. Another experiment showed that teenagers were more likely to seek out these products if they were shown content in their news feed that could be considered depressing or likely to cause anxiety.

Key Challenges

Possibly the biggest challenge is balancing the consumer's desire for personalized marketing, products, and services with the ongoing public distrust and distaste of wide-scale data gathering and behavioral analytics.

This is a hurdle that any business dealing with collecting personal data will have to find its own ways to overcome. But as regulators begin to catch up with the data harvesters and marketers, the public's confidence when it comes to trusting businesses to behave responsibly with their data should increase. With hefty fines now a threat in many jurisdictions for companies that misuse data, gather it without consent, or fail to adequately protect it, the benefits of targeted marketing and personalization could become more appealing and the pitfalls less scary. Around the world, public attitudes towards personal data marketing differ wildly but there are signs that attitudes are converging – one survey of US consumers this year found that 62% were in favor of the introduction of regulations equivalent to the European GDPR.[12]

Just as poorly targeted marketing is a turn-off for customers, so too are overly personal or familiar communications that can appear creepy in the way they are personalized. This means concerted efforts need to be made to demonstrate to customers that your business can be trusted. Clearly explaining what data is collected and how it is used, as well as offering opportunities for privacy-conscious individuals to opt out of data capture and use your products and services anonymously, are both vital strategies. Many providers, such as Google and Facebook, automate the process of allowing users to check what information is held about them, and provide tools and settings for limiting the amount or type of data that is held.

Beyond that, there are challenges inherent in the process of mass personalization and identifying micro-moments. Each trend requires the development of a mature data and analytics strategy in order to distinguish which indicators are truly valuable and what is simply "noise."

Once that's in place, the next challenge is to overcome the friction that is inevitably generated by adding an extra personalization step to the process of providing a product or service. The examples of mass personalization in practice that are given in the previous section use a variety of methods for allowing customers to submit their personalized requirements. The best ones try to make the process of submitting personal choices a pleasing and satisfying experience, using apps and web portals to give an insight into how the finished product is coming along during the manufacturing process. However, this also means that companies have to make sure their back office, manufacturing, and order fulfillment processes are optimized to deliver the customized experience.

How to Prepare for This Trend

One thing that any organization should treat as an ongoing priority process is reviewing the methods of data capture that are used, and how details about it are communicated to customers. Building trust is essential and appearing transparent and thorough in your privacy safeguarding can play a big part in this.

Once you're certain you aren't targeting customers in a way that's going to freak them out, you can think about building or utilizing tools to better understand their behavior and needs. Plenty of ready-made ones exist already, like Facebook's and Google's targeted marketing programs that leverage the vast troves of data those organizations already own.

These tools are available to everyone though, so if you need to stand out from the crowd you may need to look into more bespoke offerings that cater to your niche, or even create your own data gathering and analytics framework. This can be a costly process, so as always, make sure that your methods are in line with your strategy, and that your projects are working towards meeting overall business goals.

Moving towards personalization will often also require a root-and-branch review of logistics and supply chain processes. Producing goods and services to individual customer specifications means taking a different approach to warehousing and inventory management, and adopting smart ways of making sure you always have the correct inventory in stock. This is because design decisions may need to be made immediately prior to the product or service being delivered. When managed well, this can have a positive impact across the organization, from cutting down on the expense of storing inventory for long periods of time, to reducing waste due to spoilage.

Emerging technologies such as artificial intelligence (Trend 1) can help with this, and investigating how they might help you make better predictions around supply and demand would certainly be a forward-thinking step.

Mass personalization isn't for every business, but considering that just a few years ago the concepts of "mass production" and "bespoke design" were considered to be entirely mutually exclusive, and are now becoming increasingly common, it might be time to rethink whether this is a trend your organization should be taking part in.

Notes

1. Made to Order: The Rise of Mass Personalisation: www2.deloitte. com/content/dam/Deloitte/ch/Documents/consumer-business/ch-en-consumer-business-made-to-order-consumer-review.pdf
2. How Target Figured Out A Teen Girl Was Pregnant Before Her Father Did, *Forbes*: www.forbes.com/sites/kashmirhill/2012/02/16/how-target-figured-out-a-teen-girl-was-pregnant-before-her-father-did/#4744d10 f6668
3. Facebook helped advertisers target teens who feel "worthless": https://arstechnica.com/information-technology/2017/05/facebook-helped-advertisers-target-teens-who-feel-worthless/#
4. Balancing the See-Saw of Privacy and Personalization: The Challenges Around Marketing for Micro-Moments: https://medium.com/@petesena/balancing-the-see-saw-of-privacy-and-personalization-the-challenges-around-marketing-for-micro-1fedc9144f62
5. Google's Personalised Search Explained: https://www.link-assistant. com/news/personalized-search.html
6. Amazon Wants to Use Predictive Analytics to Offer Anticipatory Shipping: https://www.smartdatacollective.com/amazon-wants-predictive-analytics-offer-anticipatory-shipping/

7. How beverages maker Paperboat is using analytics to personalize consumer tastes: www.techcircle.in/2018/10/15/how-beverages-maker-paperboat-is-using-analytics-to-personalise-consumer-tastes

8. Custom fit earphones: Audio nirvana or a waste of money?: https://arstechnica.com/gadgets/2017/07/custom-fit-earphones-snugsue18-review/

9. Toutiao, a Chinese news app that's making headlines, *The Economist*: www.economist.com/business/2017/11/18/toutiao-a-chinese-news-app-thats-making-headlines

10. Cutting-Edge Beauty Brands Like Skinmade Are Redefining Customization With AI Technology: www.psfk.com/2019/06/reinventing-beauty-experiences-enhanced-customization.html

11. 10 Examples Of Personalization In Healthcare, *Forbes*: www.forbes.com/sites/blakemorgan/2018/10/22/10-examples-of-personalization-in-healthcare/#32ffece824e0

12. The pitfalls of personalisation: https://gdpr.report/news/2019/04/10/the-pitfalls-of-personalisation./

TREND 24
3D AND 4D PRINTING AND ADDITIVE MANUFACTURING

The One-Sentence Definition

3D printing (also known as additive manufacturing) means creating a 3D object from a digital file by building it layer by layer; 4D printing is based on the same process but with a twist – namely, a built-in ability for the printed object to transform itself.

What Is 3D and 4D Printing and Additive Manufacturing?

If there's one theme that crops up repeatedly throughout this book, it's the rise of automation. 3D printing may seem distinctly more low-tech than trends like artificial intelligence (AI) or facial recognition, but it still ties into that theme of business processes becoming more streamlined and automatic. For example, using 3D printing the factories of the future could quickly print spare parts for machinery on-site, without having to wait for those parts to be shipped half-way around the world. Even entire assembly lines could be replaced with 3D printers.

As you can imagine, 3D printing has the potential to transform manufacturing. But, as we'll see in this chapter, 3D printing has much wider applications – from the good (such as printing human tissue for transplants) to the not-so-good (printing weapons) to the that's-going-to-take-some-getting-used-to (printing food).

How does it work, though? Traditional manufacturing tends to be a subtractive process, meaning an object is typically cut or hollowed out of its source material (plastic, say, or metal) using something like a cutting tool. However, 3D printing is an *additive* process (hence the name *additive manufacturing*), which involves creating the object by adding layers upon layers of material, building up until you have the finished object. In other words, you start from nothing and build the object up bit by bit, as opposed to starting with a block of material and cutting or shaping it down into something. If you were to slice a finished 3D printed object open, you'd be able to see each of the thin layers, a bit like rings in a tree trunk.

But let's take a step back from that. Before printing anything, you need a 3D model of the object you're trying to create – a digital blueprint, if you will. That blueprint or model is then "sliced," essentially dividing the model into hundreds (or potentially thousands) of layers. This information is fed to the 3D printer and, hey presto, it prints the object slice on top of slice on top of slice…

The main benefit of 3D printing is that even complex shapes can be created much more easily, and using fewer materials than traditional manufacturing methods (good for the environment and the bottom line). Transport needs are reduced, since parts and products can be printed on-site, rather than having to order and wait for components. And one-off items can be made quickly and easily, without worrying about economies of scale – which could be a game-changer for rapid prototyping, custom manufacturing, and creating highly personalized products. What's more, the materials used for 3D printing can be pretty much anything: plastic, obviously, but also metal, powder, concrete, liquid, even chocolate.

It's no wonder, then, that the International Data Corporation predicts that worldwide spending on 3D printing will continue to grow, reaching $23 billion in 2022, up from $14 billion in 2019.[1] This acceleration of 3D printing will no doubt be driven by faster, more affordable 3D printers, and greater integration with advances like AI (Trend 1), the Internet of Things (Trend 2), voice interfaces (Trend 11), and machine co-creativity and generative design (Trend 17). To put it another way, 3D printing will become smarter, more connected, and more accessible.

What about 4D printing? 4D printing is the cutting edge of additive manufacturing. It's based on the same additive approach as 3D printing, where a 3D product or object is built up in successive layers. But, with 4D printing, the object being created can be programmed to change its shape when prompted by certain conditions or triggers (e.g. water or heat). It's 3D printing, then, but with the added dimension of transformation. For example, a storage carton could flatten itself when prompted. Furniture could assemble itself (take that, Ikea!). Structures could repair themselves after weather damage. The possibilities are endless. 4D printing is still very much in the experimental stage and we don't yet understand all the possible applications, but it certainly promises to revolutionize the field of additive manufacturing.

How Is 3D and 4D Printing and Additive Manufacturing Used in Practice?

Let's look at some of the ways 3D printing and (to a lesser extent) 4D printing are beginning to make an impact across various different sectors.

3D Printing in Manufacturing

With 3D printing, companies can create mechanical parts for easy repairs, transform production processes, allow for greater customization of products, create prototypes faster, and more.

- As one of the world's biggest manufacturers, **GE** has invested heavily in 3D printing. In fact, the company has splashed out $1.5 billion on the technology. In one example, GE is 3D printing fuel nozzles for LEAP jet engines, and is expecting to be producing 35,000 nozzles a year.[2]

- German sportswear giant **Adidas** says it can cut the time it takes to bring a new shoe design to market to just one week, thanks to 3D printing technology. The company is already 3D printing trainer soles, initially at two highly automated factories in Germany and the US, and now at some of their Chinese sites, too.[3]

- 3D printing is really catching on in the car manufacturing business. Three out of four automotive companies in Germany and the US – including companies like **BMW** and **Ford** – are using 3D printing to mass produce car components and spare parts.[4]

- **Siemens Mobility** turned to 3D printing in order to produce customized train parts on demand, including armrests for train driver seats. The company was able to create custom parts at low cost and cut the production time from weeks down to days, and has since expanded its offering of 3D printed custom parts to an on-demand online platform.[5]

3D Printing Human Tissue

It might surprise you to learn that the health sector was one of the early adopters of 3D printing technology. For example, thanks to 3D printing, it's possible to create prosthetics that can be easily tailored to each individual's body type and needs. But the medical uses of 3D printing extend far beyond prosthetics:

- At the **Wake Forest Institute for Regenerative Medicine**, researchers have been able to print bones, muscles, and ears – known as *bioprinting* – and implant them successfully into animals.[6] What's key is that the printed tissue survived after being implanted and became functional tissue.

- The ability to bioprint whole, functioning organs is still some way off, but scientists have been able to print organ tissue. For example, at the **MRC Centre for Regenerative Medicine** at the University of Edinburgh, scientists have successfully printed liver cells that can, so far, stay alive for up to a year.[7] The hope is technology like this will, in the long term, help to provide liver support for patients with chronic liver disease.

- In an experiment at the **Northwestern University Feinberg School of Medicine in Chicago**, a mouse was implanted with synthetic, printed ovaries. The mouse went on to give birth to healthy babies.[8]

Printing Food

Pretty much any material can be 3D printed, so why not food?

- **Choc Edge** sells 3D printers that allow chocolatiers to design and produce amazingly inventive chocolates in pretty much any shape or design.[9] As with any 3D printing process, the shape is sliced into layers, which are then built up by the printers in one ultra-thin layer of melted chocolate at a time. The chocolate cools and sets as it's printed. Hershey has also been experimenting with 3D printed chocolate.[10]

- Sticking with the sweet theme, Ukrainian architect-turned-pastry-chef **Dinara Kasko** has made a name for herself on

Instagram by posting pictures of her striking geometric 3D printed pastries.[11]

- Startup **Novameat** claims it has printed the world's first 3D printed vegan steak, made from plant-based proteins. According to the company, the meat-free steak successfully mimics the fibrous, fleshy nature of meat, but is far more sustainable than animal agriculture.[12] With a burgeoning world population (set to hit more than 9 billion by 2050), the race is on to create a sustainable food supply that can feed people without destroying the planet, and it looks like 3D printing could be part of the solution.

Printing Buildings

Architecture and construction are also being enhanced by 3D printing technology. Could this solve the problem of providing affordable housing? These examples show what's possible.

- Russian startup **Apis Cor** is able to 3D print a modest house in just 24 hours, and save up to 40% on construction costs.[13] What's more, thanks to the mobile nature of Apis Cor's printing devices, houses can be printed on-site rather than in a factory. The mobile printer lays down layers of a concrete mixture to build up the walls, then, once the printer is removed, insulation, windows, and a roof are added. In the next phase of its development, the company is working to create equipment for 3D printing foundations, floors, and roofing for high-rise construction.

- **Dubai** has set an ambitious goal of 3D printing 25% of buildings by 2030, and is working with 3D printing construction firm Cazza to achieve that aim.[14] Using 3D printing robots, the company is planning to create new large-scale developments of low-rise buildings in Dubai.

- San Francisco housing non-profit New Story collaborated with construction technology company **Icon** to produce a tiny house that cost $10,000 dollars and took just 48 hours to build – and that was with the printer running at just 25% speed.[15] Based on this, Icon reckons it could now build a 600–800-square-foot house in just 24 hours for $4,000.

4D Printing in Action

It's early days for 4D printing, but these examples give a glimpse of what might be possible in the future.

- The **MIT Self-assembly Lab** is dedicated to inventing self-assembly and programmable material technologies. In one example, a flat-printed structure slowly folds itself into a cube shape once placed in hot water.[16] This sort of technology could have wide implications for manufacturing, construction, production assembly, and more.

- French company **Poietis** claims it can print tissue cells that can "evolve in a controlled way."[17]

- Researchers at the **Lawrence Livermore National Laboratory** have printed silicone material that is flexible and can adapt itself when heat is applied. This could, for example, be used to create shoes that can grow and expand as the wearer grows – truly customizable shoes, in other words.[18]

Key Challenges

As with all of the trends in this book, 3D printing technology brings many opportunities – but it also brings some downsides, challenges, and obstacles to overcome.

While 3D printing has the potential to reduce the environmental impact of manufacturing (by using fewer materials overall), we have to consider the environmental impact of the printers themselves. For one thing, 3D printing tends to rely on plastic – although that will change as 3D printing in metal, concrete, and other materials becomes more commonplace. But perhaps the main concern is that 3D printers use a lot of energy – potentially hundreds of times more energy than traditional methods such as molding, casting, or machining.[19]

3D printing also presents problems for intellectual property owners, since the technology enables counterfeiters to produce fake licensed goods cheaply and easily (counterfeit Star Wars toys, for instance). Such digital piracy could amount to a loss of $100 billion a year in IP globally, according to Gartner.[20]

There's also the problem that weapons can be easily 3D printed. In 2019, a British student was convicted of manufacturing a gun using 3D printing – believed to be the first conviction of its kind in the UK. The man claimed the gun was for a dystopian film project, but couldn't explain why he had printed a fully working, potentially lethal design instead of a fake gun.[21]

3D printing may also have implications for worker safety, as some research has indicated additive manufacturing methods have the potential to cause injuries to workers.[22] In particular, workers may be exposed to ultrafine metal and other particles created during the additive manufacturing process, which could lead to health problems further down the line. The science is still evolving in this field, but it's certainly something for risk managers to consider. Manufacturers will have to assess the materials they use in additive manufacturing, and look at factors such as ventilation and the proper disposal of residue materials.

How to Prepare for This Trend

At the time of writing, 3D printing is far from commonplace, but as the examples in this chapter have shown, the technology has the potential to challenge traditional production methods. Therefore, if your business involves manufacturing products or components of any kind, you'll want to consider how 3D printing could augment your manufacturing operations.

One thing that I find particularly exciting about 3D printing is the potential it brings for mass personalization of products (see Trend 23 for more on mass personalization). Thanks to 3D printing, products and designs can be customized to suit one-off requests and orders – and this could cover anything from personalized sneakers to food that's customized to our individual nutritional needs.

As consumers, we've all grown comfortably accustomed to having products and services personalized to our needs. The smart thermostat that regulates the temperature of your home according to how you use the space. The TV streaming platform that understands what you like to watch and serves up more of the same content. The fitness tracker that helps you achieve your unique health and fitness goals. Giving customers exactly what they want is a key ingredient for business success. But customizing products has traditionally been an expensive and labor-intensive process. 3D (and 4D) printing has the potential to change all that. While some remain skeptical about the widespread adoption of 3D printing, I believe this scope for increased personalization will make all the difference for the future of 3D printing. Therefore, if you believe that your customers would welcome more personalized products, it might be worth considering 3D printing as a means to achieving that.

Notes

1. IDC Forecasts Worldwide Spending on 3D Printing to Reach $23 Billion in 2022: www.idc.com/getdoc.jsp?containerId=prUS44194418
2. 3D printers start to build factories of the future, *The Economist*: www.economist.com/briefing/2017/06/29/3d-printers-start-to-build-factories-of-the-future

3. 3D printers start to build factories of the future, *The Economist*: www.economist.com/briefing/2017/06/29/3d-printers-start-to-build-factories-of-the-future

4. Start Your Own 3D Printing Business: 11 Interesting Cases Of Companies Using 3D Printing: https://interestingengineering.com/start-your-own-3d-printing-business-11-interesting-cases-of-companies-using-3d-printing

5. Siemens Mobility Overcomes Time and Cost Barriers of Traditional Low Volume Production for German Rail Industry with Stratasys 3D Printing: http://investors.stratasys.com/news-releases/news-release-details/siemens-mobility-overcomes-time-and-cost-barriers-traditional

6. Wake Forest Researchers Successfully Implant Living, Functional, 3D Printed Human Tissue Into Animals: https://3dprint.com/119885/wake-forest-3d-printed-tissue/

7. Liver success holds promise of 3D organ printing, *Financial Times*: www.ft.com/content/67e3ab88-f56f-11e7-a4c9-bbdefa4f210b

8. 3D-Printed Ovaries Offer Promise as Infertility Treatment: www.livescience.com/59189-3d-printed-ovaries-offer-promise-as-infertility-treatment.html

9. Choc Edge: http://chocedge.com/

10. You can now 3D print complex chocolate structures, *Wired*: www.wired.co.uk/article/cocojet-chocolate-3d-printer

11. We Interviewed Dinara Kasko: 3D Printing Instagram Food Sensation: www.3dnatives.com/en/dinara-kasko-pastry-chef060420174/

12. Novameat develops 3D-printed vegan steak from plant-based proteins: www.dezeen.com/2018/11/30/novameat-3d-printed-meat-free-steak/

13. #3DStartup: Apis Cor, Creators of the 3D printed house: www.3dnatives.com/en/apis-cor-3d-printed-house-060320184/

14. This Startup Is Disrupting The Construction Industry With 3D-Printing Robots, *Forbes*: www.forbes.com/sites/suparnadutt/2017/06/14/this-startup-is-ready-with-3d-printing-robots-to-build-your-house-fast-and-cheap/#25aa3d016e8e

15. These 3D-printed homes can be built for less than $4,000 in just 24 hours: https://www.businessinsider.com/3d-homes-that-take-24-hours-and-less-than-4000-to-print-2018-9?r=US&IR=T

16. MIT Self-assembly Lab: https://selfassemblylab.mit.edu/

17. Four Ways 4D Printing is Becoming a Reality: www.engineering.com/3DPrinting/3DPrintingArticles/ArticleID/18551/Four-Ways-4D-Printing-Is-Becoming-a-Reality.aspx

18. Lab researchers achieve "4D printed" material: www.llnl.gov/news/lab-researchers-achieve-4d-printed-material
19. The dark side of 3D printing: 10 things to watch: www.techrepublic.com/article/the-dark-side-of-3d-printing-10-things-to-watch/
20. Gartner Says Uses of 3D Printing Will Ignite Major Debate on Ethics and Regulation: www.gartner.com/en/newsroom/press-releases/2014-01-29-gartner-says-uses-of-3d-printing-will-ignite-major-debate-on-ethics-and-regulation
21. UK student convicted for 3D printing gun: https://futurism.com/the-byte/uk-student-convicted-3d-printing-gun
22. Tackling the risks of 3D printing: www.aig.co.uk/insights/tackling-risks-3d-printing

TREND 25
NANOTECHNOLOGY AND MATERIALS SCIENCE

The One-Sentence Definition

Nanotechnology essentially means controlling matter on a tiny scale, at the atomic and molecular level, while materials science is the study of materials – characteristics, properties, uses, and so on – to understand how various factors influence a material's structure.

What Is Nanotechnology and Materials Science?

Here, the two concepts are combined into one chapter since both nanotechnology and materials science are giving us exciting new materials and products – including tiny chips and sensors, bendable displays, longer-lasting batteries, and even lab-grown food. In time, we can expect advances in nanotechnology and materials science to feed into other trends already discussed in this book, such as smart devices (Trend 2), smart cities (Trend 5), autonomous vehicles and drones (Trends 14 and 19), gene editing (Trend 16), and 3D and 4D printing (Trend 24).

Biotechnology – which applies biological processes to industrial purposes – is closely related to both trends, and is leading to breakthroughs such as lab-grown human tissue. In this chapter, I'll largely focus on nanotechnology and materials science, with a few examples from the world of biotechnology.

Let's start with a quick overview of nanotechnology. Everything around you, from the chair you might be sitting in, to the book or tablet you're holding in your hand, is made of atoms and molecules (which are atoms linked together). Nanotechnology is about looking at the world on such a tiny scale that we can not only see the atoms that make up everything around us (including ourselves), but we can manipulate and move those atoms around to create new things. In this way, nanotechnology is a bit like construction, but on a tiny scale.

How tiny? Forget microscopic. We're talking *nanoscopic*. The nanoscale is 1,000 times smaller than the microscopic level and a billion times smaller than the typical world of meters and kilometers that we're used to measuring things in. (Nano literally means one billionth.) If you took a human hair, for instance, it would measure approximately 100,000 nanometers wide. A strand of human DNA is just 2.5 nanometers wide. That's the sort of scale we're talking about.

Why does the nanoscale matter? Because, when we look at objects and materials on an atomic and molecular level, we can understand more about how the world works. There's also the fact that certain substances behave differently and have completely different properties at an atomic level. Think of how a diamond and the graphite in a pencil are both made from carbon; when the carbon atoms bond a certain way, you get a diamond, and when they bond another way, you get graphite.

In another example, silk may feel incredibly soft and delicate to the touch, but zoom in to a nano level and you'd see it's made up of molecules aligned in cross-links, which is what makes it so strong. We can then use knowledge like this to manipulate other materials at a nano level to create super-strong, state-of-the-art materials like Kevlar. Or products that are lighter. Or fabrics that are stain resistant.

This is where the *technology* bit of nanotechnology comes in – using our knowledge of materials at a nano level to create new solutions.

You can see, then, how nanotechnology and materials science are linked. The study of materials at a nano level could be considered almost a subfield of materials science, where the materials are being looked at on an atomic and molecular level. However, nanotechnology also incorporates principles from other areas of science, such as molecular biology and quantum physics, which is why it's usually treated as a separate discipline.

Today, tiny computer chips, transistors, and smart phone displays are all being built using nanotechnology and materials science. But the really exciting advances are perhaps still decades away – advances like *nanomachines* and *nanobots*, which could be injected into the human body to perform cellular repairs, or *hypersurfaces*, where any surface could be transformed into a touch screen interface. In theory, if we can manipulate atoms, we can create pretty much anything.

How Is Nanotechnology and Materials Science Used in Practice?

Let's look at some fascinating examples across the fields of nanotechnology, materials science, and biotechnology.

Nanotechnology in Manufacturing

Many of the practical applications of nanotechnology are seen in manufacturing, where the technology is being used to create innovative products that are stronger, lighter, and more durable – products that perform better, in other words.

- **MesoCoat** has developed a nanocomposite coating called CermaClad, which is designed to coat pipes used in the oil industry to make the pipes corrosion and wear resistant.[1]

- By coating the foam used in **upholstered furniture** with carbon nanofibers, manufacturers can reduce flammability by up to 35%.[2]

- In **tennis**, nanotechnology helps tennis balls keep their bounce for longer, and makes tennis racquets stronger.[3]

- **Nanorepel** makes high-performance nanocoatings that can be used to protect your car's paintwork from bird droppings.[4]

- Nanotechnology powers many of the electronics we use in everyday life, with **Intel's** tiny computer processors being one example. The latest generation of Intel's Core processor technology is an impressive 10-nanometer chip.[5]

Advances in Materials Science

Let's look at just a few of the developments leading the way in materials science, many of which incorporate advances from nanotechnology.

- Thanks to the development of carbon fibers – fibers composed of carbon atoms – we now have composite materials that are incredibly strong, light, and high performance. The **Boeing** 787 Dreamliner uses such composites in its fuselage and wings.

- At just one atom thick, **graphene** is the thinnest material in the world. It's amazingly strong – 200 times stronger than steel – yet it's flexible and can be bent into different shapes. When added to other materials, such as ceramics and metals, graphene has the potential to make them stronger, more flexible, and resistant to rust or corrosion. Think of bendable solar cells, rust-free metal coatings, and anti-corrosion paint…

- **HyperSurfaces** is developing technology that can turn any object, surface, or material into an intelligent surface that can detect motion and carry out commands. For example, your coffee table could become the controller for your TV, lighting, and thermostat. The startup has reportedly received a lot of interest from car manufacturers.[6]

- The portable electronic devices we own are possible thanks to the development of lithium ion batteries (or Li ion batteries), which are relatively small and light and have a high energy density. But researchers are racing to make batteries better: smaller, higher energy, longer life, and less damaging to the environment. Improving battery performance is particularly important if we're to store green energy and increase adoption of electric vehicles. Scientists at **Toyota** have been testing materials for a battery that can fully charge or discharge in just seven minutes – ideal for electric cars.[7] Similarly, **Grabat** has created graphene batteries that can charge and discharge 33 times faster than Li ion batteries, and give electric cars a driving range of up to 500 miles on one charge.[8] There's even a foldable battery, the Jenax J. Flex battery, which could pave the way for the bendable gadgets of the future.[9]

Smart Materials and Self-Healing Materials

What else can we expect of the materials of the future? If these examples are anything to go by, manufacturers will increasingly turn to materials that can change their properties or repair themselves automatically.

- Inspired by the human body's ability to heal, scientists are working to create **self-healing materials** that could repair damage or wear by themselves. The first commercially available self-healing materials are likely to be paints and coatings that fix themselves when they get stained or weather damaged.[10] But in time, who knows? Maybe we'll have bridges that can repair themselves when cracks appear.

- Smart materials have properties that change according to their surroundings. A common example of this is the **photochromic lenses** used in glasses, which turn into sunglasses when exposed to sunlight, then return to normal glasses when indoors.

- **Shape memory polymer** – a material that can be bent back into its original shape when heated – is another example of a smart material. According to one patent, this technology could be used to make car bumpers easier to fix after accidents. So, when you ding your car bumper in future, in theory it could be easily returned to its original shape.[11]

Biotechnology in Action

Biotechnology uses biological systems to develop new technologies across all sorts of industries, including healthcare, manufacturing, and agriculture. Medical biotechnology, for example, has given us advances such as vaccines and antibiotics.

- In her book *The Age of Living Machines*, MIT President Susan Hockfield predicts a future in which "nature's genius" is leveraged to solve some of humanity's biggest challenges. Essentially, Hockfield believes biology and engineering will converge to produce technologies we can't yet conceive.[12]

- In agriculture, biotechnology is what gives us **genetically modified crops**, which help producers increase crop yields or grow crops capable of resisting pests and diseases. Biotechnology has also led to additional nutrients being added to foods, such as "golden rice," which is infused with beta carotene. For more on genomics and gene editing see Trend 16.

- A team from **Duke University** in North Carolina has developed a patch to replace heart muscle cells destroyed by a heart attack. This essentially involves growing pieces of heart muscle in a lab to create a patch that could be surgically attached to patients after a heart attack. The patch has been tested successfully in rodents.[13] Read more about the physical augmentation of humans in Trend 3.

- We're not just growing human tissue in labs. In an effort to create a more sustainable and ethical food supply, researchers have been experimenting with growing food in labs. Lab-grown food, or cellular agriculture, involves taking cells from an animal (e.g. a cow or chicken) and placing it in a growing medium in a bioreactor to produce "cultured meat." The same is true of seafood. For example, **BlueNalu** works to extract muscle cells from various seafood to be cultivated in a lab. You might also like to revisit Trend 24, which touches on 3D printing food.

Key Challenges

The manipulation of materials on a nanoscale has prompted some significant concerns; specifically, that we don't know what effect nanoscale machines or organisms might have on the environment or the human body. We know that tiny particles can cause great harm to the body – consider, for example, the wide use of certain chemicals and materials in previous decades that have since been proven toxic to humans. Might nanomaterials pose a similar threat? After all, they would be small enough to penetrate the blood–brain barrier, which protects the brain from foreign substances. If we end up using nanoparticles in everything from clothing to sunscreen to pipework, how will we know those particles won't ultimately poison us?

The "grey goo" scenario is the most commonly mentioned nightmare outcome of nanotechnology. According to this theory, humans could create dangerous replicating nanobots that basically eat up everything in the biosphere in an unstoppable attempt to replicate themselves, destroying everything in their wake. It may sound far-fetched – and

Dr K. Eric Drexler, the nanotechnology pioneer who coined the phrase, later said he wished he'd never coined the phrase "grey goo" – but the underlying principle is compellingly simple: what if humans inadvertently create something that's harmful to human life, rather than beneficial? (Of course, humans have a long history of doing just that – cigarettes and nuclear weapons being just two examples.)

Assuming we don't accidentally annihilate all existing lifeforms, there may be other problems to contend with. Let's say nanobots help to eliminate all disease, so humans live far longer. What impact will that have on the planet? Should we even want to augment humans to such a degree that we end up no longer, well, human? (Read about the notion of *transhumans* in Trend 3.)

There's also the concern that nanotechnology could be used by criminals and terrorists. For one thing, it may be possible to create miniscule weapons that are almost impossible to detect. Ultimately, the benefits of nanotechnology will have to outweigh potential risks like this. Many in the field believe they will.

How to Prepare for This Trend

A lot of the cutting-edge work in nanotechnology and materials science is being done in academic institutions, so we're yet to see these spread widely into the world of business. But, in the future, a vast array of companies may benefit from nanotechnology, particularly those in the manufacturing industry. If you think of the potential to create products that are stronger, lighter, safer, and smarter, it's clear how nanotechnology and materials science may help to deliver a huge competitive advantage.

Depending on your industry, this is likely to be a trend that you keep an eye on, rather than rush to develop a nanotechnology strategy. But as the technology evolves, and you begin to consider the applications for your business, you might like to keep in mind questions such as:

- Is there a compelling business reason to apply nanotechnology? For example, does nanotechnology offer a way to increase the performance of your products? As with any of the trends in this book, technology for technology's sake is rarely a good idea.

- What are the safety implications for your business? In other words, how will you make sure nanoparticles are safe for customers? And how will you protect employees working with nanotechnology products?

- What are the environmental risks? Consider also the sustainability of nanomaterials.

Notes

1. MesoCoat Receives Two (New) Grants to Develop CermaClad Arc Lamp Applications: www.businesswire.com/news/home/2014100 6005918/en/MesoCoat-Receives-New-Grants-Develop-CermaClad% E2%84%A2-Arc
2. Carbon Nanofibers Cut Flammability of Upholstered Furniture: www.nist.gov/news-events/news/2008/12/carbon-nanofibers-cut-flammability-upholstered-furniture
3. Nanotechnology in sports equipment: The game changer: www.nanowerk.com/spotlight/spotid=30661.php
4. Nanorepel: www.nanorepel.eu/?lang=en
5. Intel's New 10-Nanometer Chips Have Finally Arrived, Wired: www.wired.com/story/intel-ice-lake-10-nanometer-processor/
6. HyperSurfaces turns any surface into a user interface using vibration sensors and AI, Techcrunch: https://techcrunch.com/2018/11/20/hypersurfaces/
7. Future batteries, coming soon: Charge in seconds, last months and charge over the air: www.pocket-lint.com/gadgets/news/130380-future-batteries-coming-soon-charge-in-seconds-last-months-and-power-over-the-air

8. Future batteries, coming soon: Charge in seconds, last months and charge over the air: www.pocket-lint.com/gadgets/news/130380-future-batteries-coming-soon-charge-in-seconds-last-months-and-power-over-the-air

9. Jenax J. Flex battery: https://jenaxinc.com/

10. Self-healing materials: www.explainthatstuff.com/self-healing-materials.html

11. Automobile bumper based on shape memory material: https://patents.google.com/patent/CN101590835A/en

12. Susan Hockfield on a new age of living machines: http://news.mit.edu/2019/3q-susan-hockfield-new-age-living-machines-0507

13. Lab-grown patch of heart muscle and other cells could fix ailing hearts, *Science*: www.sciencemag.org/news/2019/04/lab-grown-patch-heart-muscle-and-other-cells-could-fix-ailing-hearts

A FEW FINAL WORDS

I hope that you enjoyed this book and have gained a better understanding of the 25 tech trends that are shaping this fourth industrial revolution. Maybe, like me, you also find many of the tech trends exhilarating and terrifying in equal measures. I also hope you got a sense of how transformative these technologies are and how much disruption and opportunities this fourth industrial revolution will bring. Many of these technologies by themselves would have a massive impact on business and society, but collectively the change will be beyond what many of us can imagine today. They will augment our jobs, transform business models, and redefine businesses and industries.

As with all the previous industrial revolutions, there will be winners and losers. It is our responsibility to manage the transition and it is up to us to make sure we use these technologies in a way that creates a better world for all of us. We must make sure these technologies serve us as humans, make our lives better, and help us solve some of the greatest challenges we face. We must ensure we use these powerful tools for good.

We have never had more powerful technologies at our disposal to help us solve the biggest challenges we as humans face. We can use these amazing technologies to tackle climate change, eliminate hunger, reduce inequalities and poverty, tackle disinformation and fake news, improve access to great healthcare, and make our cities and societies more resilient and sustainable. Let's make it happen!

I have the pleasure to work with so many wonderful businesses and governments every day that want to better understand and use future technologies, and I am optimistic that the vast majority of them are going to use these technologies to make our world better, and more human.

I would love to establish a dialogue beyond the confines of this book. Let me know if you have any questions, feel free to share any success stories, and get in touch if you feel that I could help your business with leveraging future technologies. Please feel free to connect with me on the following platforms:

LinkedIn: Bernard Marr
Twitter: @bernardmarr
YouTube: Bernard Marr
Instagram: @bernardmarr
Facebook: facebook.com/BernardWMarr

Or head for my website at **www.bernardmarr.com** for more content as well as an opportunity to join my weekly newsletter, in which I share the very latest information.

ABOUT THE AUTHOR

Bernard Marr is an internationally best-selling author, popular keynote speaker, futurist, and strategic business and technology advisor to governments and companies. He helps organizations and their management teams prepare for a new industrial revolution that is fueled by transformative technologies like artificial intelligence, big data, blockchains, and the Internet of Things.

Bernard is a regular contributor to the World Economic Forum, writes a weekly column for *Forbes*, and is a major social media influencer, with his LinkedIn ranking among the top five in the world and number one in the United Kingdom. His 1.5 million followers on LinkedIn and strong presence on Facebook, Twitter, YouTube, and Instagram give him a platform that allows Bernard to actively engage with millions of people every day.

Bernard has written over 15 books and hundreds of high-profile reports and articles, including the international best-sellers Artificial Intelligence in Practice, Big Data in Practice, Data Strategy, The Intelligent Company and The Intelligence Revolution.

Bernard has worked with or advised many of the world's best-known organizations, including IBM, Microsoft, Google, Walmart, Shell, Cisco, HSBC, Toyota, Vodafone, T-Mobile, the NHS, Walgreens Boots Alliance, the Home Office, the Ministry of Defence, NATO, the United Nations, among many others.

ABOUT THE AUTHOR

Connect with Bernard on LinkedIn, Twitter (@bernardmarr), Facebook, Instagram, and YouTube to take part in an ongoing conversation and head to www.bernardmarr.com for more information and hundreds of free articles, white papers, and e-books.

If you would like to talk to Bernard about any advisory work, speaking engagements, or influencer services, please contact him via email at hello@bernardmarr.com

ACKNOWLEDGMENTS

I feel extremely lucky to work in a field that is so innovative and fast moving and I feel privileged that I am able to work with companies and government organizations across all sectors and industries on new and better ways to use the latest technology to deliver real value – this work allows me to learn every day and a book like this wouldn't have been possible without it.

I would like to acknowledge the many people who have helped me get to where I am today. All the great individuals in the companies I have worked with who put their trust in me to help them and in return gave me so much new knowledge and experience. I must also thank everyone who has shared their thinking with me, either in person, blogposts, books, or any other formats. Thank you for generously sharing all the material I absorb every day! I am also lucky enough to personally know many of the key thinkers and thought leaders in the field and I hope you all know how much I value your inputs and our exchanges.

I would like to thank my editorial and publishing team for all your help and support. Taking any book from idea to publication is a team effort and I really appreciate your input and help – thank you Annie Knight, Kelly Labrum, and Samantha Hartley.

My biggest acknowledgment goes to my wife, Claire, and our three children, Sophia, James, and Oliver, for giving me the inspiration and space to do what I love: learning and sharing ideas that will make our world a better place.

INDEX

INDEX